Lecture Notes in Morphogenesis

Series Editor

Alessandro Sarti, CAMS Center for Mathematics, CNRS-EHESS, Paris, France
e-mail: alessandro.sarti@ehess.fr

More information about this series at http://www.springer.com/series/11247

Stefania Caliandro

Morphodynamics in Aesthetics

Essays on the Singularity of the Work of Art

Springer

Stefania Caliandro
École supérieure d'art et de design des Pyrénées
Tarbes, France

ISSN 2195-1934 ISSN 2195-1942 (electronic)
Lecture Notes in Morphogenesis
ISBN 978-3-030-29630-8 ISBN 978-3-030-29631-5 (eBook)
https://doi.org/10.1007/978-3-030-29631-5

This Springer imprint is published by the registered company Springer Nature Switzerland AG
The registered company address is: Gewerbestrasse 11, 6330 Cham, Switzerland

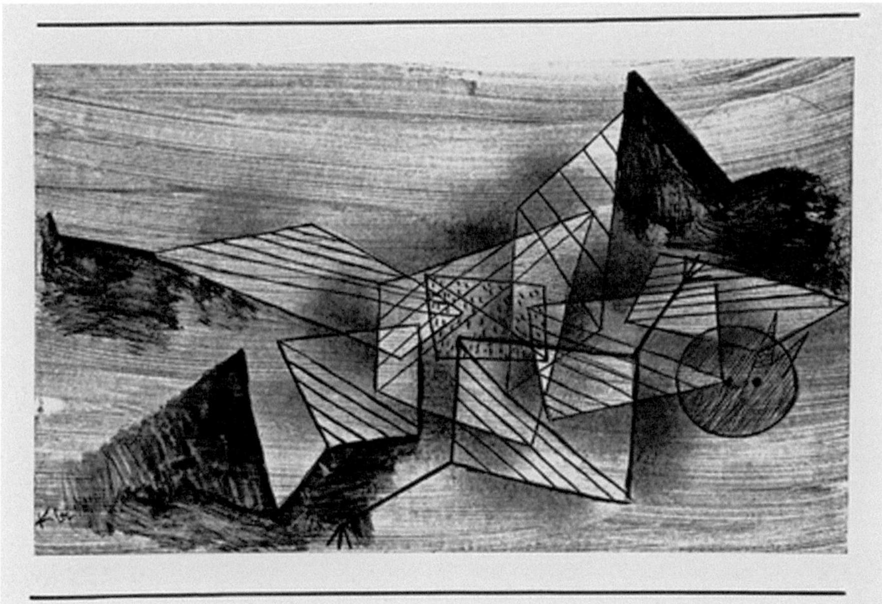

Paul Klee. *Katastrophe im Winter—Catastrophe in Winter*. 1930, 229. Watercolour and coloured paste on paper on cardboard. 21 × 33 cm. Zentrum Paul Klee, Bern. *Photo* © Zentrum Paul Klee

Contents

Chapter 1
The Singularity of the Work of Art. Enquiry for a Morphodynamic Semiotics

A renewed morphogenetic approach supported by physics and mathematics has recently sparked theoretical considerations in humanities concerning the perception and emergence of forms as well as the construction of a meaning, perceptually grounded, associated with these forms. Semiotics, which has historically worked on sense and significance, intends now to bridge social and natural sciences on this fundamental point. In dealing with morphodynamic studies, semiotics has, however, engaged itself in a progressive but deep transformation of its own foundations and included, for instance, sensorial and perceptual activities, whether conscious or not, in the act of semiosis. Analyses that were previously reserved to cognitive features in order to define the contents of an object or phenomenon are now integrating a more comprehensive methodology focusing on forms, on how forms do manifest and express themselves and become relevant to the observer's examination. In this sense, such a fertile ground is helping to rethink about a scientific theoretical approach to art, despite the extremely rich variety and unpredictability of aesthetic creations.

Resuming the project of a science of art does not mean reactivating positivist theories about art deployment or its possible explanations, but to assume the utopia of a functional[1] description that takes into account the perceptual and cultural participation of the beholder in the construction of the artistic phenomenon. Instead of delving into the work of art by means of a priori tools to detect basic meanings, the challenge is to accept and deal with the complexity of the aesthetic phenomenon wherein every secondary aspect may be potentially significant. The complexity which concerns every understanding of an ordinary object in relation with its environment is indeed particularly striking in art, in compliance with the subversive and puzzling force of art realisations. Moreover, the idea that the sense arises from the experience of the work of art may endorse a semiogenetic approach inheriting the physic orientation Jean Petitot has consistently bolstered in semiotics following René Thom's morphogenetic theory. Also the Groupe μ's latest research discredits theories of meaning

[1] Somehow, paying due regard to the unattainable idea of a comprehensive scientific description of the work of art, I would willingly opt for the qualifier *dysfonctionnal* if it were not in the sense of a semiotic function.

S. Caliandro, *Morphodynamics in Aesthetics*, Lecture Notes in Morphogenesis,
https://doi.org/10.1007/978-3-030-29631-5_1

founded on conventionality or codes and, instead of endorsing, as previously done, the study of norms and average cultural habits to measure deviations and specificities, advocates a genetic semiotics exploiting neurosciences', psychological, sociological, epistemological and cognitive contributions (cf. Groupe μ 2015, p. 20). In this way, Groupe μ is deeply reviewing traditional structuralism and has admitted that perception is not *exterior* to structure (cf. ibid., p. 32)[2] and, therefore, must be taken as a key-moment in shaping semiotic values.

It is rather hard to adumbrate a general methodology to analyse the various genres such as painting, sculpture, installation, performance, architecture, photography, video and digital art, just to limit our enquiry to the field of the so-called *visual* arts. Most of the semiotic studies focus primarily on planar surfaces, as used for drawing, image and canvas, then expanding and adapting the findings to other specific constraints. This is not without the risk of elaborating strongly oriented and refined tools that are not very adequate in other artistic fields. To avoid dangerous generalities, it could even be rightfully asserted, in my opinion, that every work of art determines its possible tools of analysis. However, since every theory has to delimit itself and take on its boundaries, I will propose to start with considerations coming from the visual field, albeit the debatable and very dubious question of visual primacy in aesthetics over other senses' appeals, for example with respect to *plasticity* or *embodiment*. Thanks to historical investigations and the large exploitation of this medium, it is indeed in the visual field that a sensorial and perceptual morphogenesis may easily be depicted. Anyhow, notwithstanding the limits of specific sensory competences, there seems to be no reason to believe that perception of forms, shapes and figures, categorisation, recognition of iconic items, the assembling and correlating arrangements such as configuration differ radically depending on the senses. These operations seemingly cross all perceptual modes, as well as their possible intermingling activation.

According to Groupe μ's *Principia semiotica*, the emergence of a form implies an act of thresholding ("*seuillage*") in the visual field. The authors acutely observe that any elementary act of knowledge results from the experience of a discontinuity: an undifferentiated field does not generate one. Thus, a threshold is a qualitative or quantitative differentiation, splitting the visual field; it induces the perception of a limit, be it sharp or fuzzy, and therefore the emergence of boundaries and forms. In this way, they identify in translocal qualities and in contrasting entities, named *dipoles*, the basic criteria organising the aesthesic apprehension. Through distinguishing and grouping, both qualities and entities lead to more and more complex semiotic operations like the abovementioned categorisation, recognition, etc. (Groupe μ 2015, pp. 76–88, esp. 80ff.). From neurological functioning to composite interpretations, these processes guarantee the elaboration of stimuli and data for human activities.

This too much abridged presentation of Groupe μ's theory allows to highlight the relevance of differentiating and assembling processes in perception, partly foreseen by the Gestalt psychological theory, as well as the experiential participation of the beholder's perceptual and cognitive system in constructing the observed object. It is

[2]On the problem of structure as related to aesthetic semiosis, please refer to Caliandro (2019).

also very valuable to consider the act of thresholding at the very morphodynamic origin of the emergence of perceptual items. As artists discerned, in particular abstract art theorists, the visual field is in fact not apprehended through a uniform continuous scanning but following tensions and saliences that the human eyes encounter in seeing. Nevertheless, the dual principle of dipoles that Groupe μ discovers in perception and in further semiotic operations might look like an unquestioned legacy of the traditional reasoning the French school has adopted in semiotics, especially from Algirdas Julien Greimas. The contrasting principle seems to clarify the bifurcation that may occur in a continuum where suddenly two coexisting entities differentiate from each other. Yet, a treatment of the perceptual information arranged in dichotomies may only be an illustrative exemplification of much richer phenomena where various different entities and qualities interact concomitantly. Bifurcation is definitively the simplest among the catastrophes examined by René Thom.[3] Moreover, even an artist like Paul Klee, balancing opposite elements to study the correspondent *weight* of forms, colours, dimensions and positions in the visual field, once the didactic discourse cleared up, feels compelled to symbolically refer to more simultaneous scales and measure their conjoint dynamism. It is in fact in the complexity of the dynamics that a work of art generally manifests itself, be it a coordinated or conflicting composition of many entities and qualities altogether.

The statement that the field is the "first visual organisable percept" (Groupe μ 2015, p. 76, own translation) should be reconsidered, in my opinion, in the sense of a simplifying idea not implying any chronological or morphological priority in the construction of the visual experience. The classical trend to put order into the logics of perception, remnant of a longstanding disposition to decompose grammatically plans and components of a work of art, reveals not only to be fallacious but precariously inconsistent with the space-time that the aesthetic phenomenon contributes to create. Similarly, as regards the limits of the artistic object, even though standard formats habitually orient the perceiver, it's the work of art itself in relation with the beholder that defines its extent, borders, matter, support and the area of the perceptual field that is addressed. The anisotropy of the visual field, noticed in psychophysics since the beginning of the twentieth century,[4] thus matches with the specific settings the work of art brings about. Likewise, any element—be it a colour, a form or a line—is not dissociable from all the relationships it has, also with respect to personal sensibility and experience. Francis Édeline, member of Groupe μ, has thoroughly analysed the semiotic potential of a line in its various features (e.g. as a limit, or as a physical concretion, i.e. a *trait* in French) and differed its values according to the framework (cf. Édeline 2008). Hubert Damisch has gone even deeper in his treatise on the line, titled *Traité du trait*, examining how articulated can be the sense of a line with respect to its sole relationship with the support, whatever it may be. The *discrimination* the line introduces, its departure from and with ("*d'avec*") the surface, determinates a

[3] Subsequent studies (i.e. Vladimir I. Arnold) have also indicated that René Thom's catastrophes cannot be reduced to his seven elementary types and that infinite variants should be included in the mathematical model. Cf. Wildgen (2017).

[4] Cf. also Schapiro (1969).

difference[5]: while distinguishing from the surface, the line does not stop being part of it and reveals to be significant with it. Deleuze (1968, p. 52; English transl.: p. 35) wrote about this difference as a *differential* in mathematical terms and associated it to a catastrophe. Now, Damisch philosophically relates the difference to Wittgenstein's thoughts on the *aspect* and to Sartre's concept of image.[6] Every element in a work of art, no matter how minimal it is, such as a line, can indeed be seen in more than one way, in more than one *aspect*, taking this term as close to the verb mode in grammar. It can arise in varied ways in the consciousness, for instance, for its material aspect or for the imaging it conveys (a representation, mental associations, etc.). As Damisch explains, the aspect is neither a thing nor a state, it exists only because it can change, involved as is in a network of determinations where no distinction is possible between seeing, imagining and interpretation, or between seeing and knowing (cf. Damisch 1995, p. 89). He then concludes, the aspect is the index of an activity, a gesture, revealing itself into the structure of the observed phenomenon; for being transitory and subject to change, the aspect points at the *structural* condition of the work of art (cf. ibid., p. 90).

As mentioned, Gilles Deleuze's *Différence et répétition* considers difference a mathematic differential. This occurs when something (e.g. a form) distinguishes itself from what (the matter, the ground) cannot distinguish itself from it, the latter continuing to espouse the former (cf. Deleuze 1968, pp. 43–95, esp. 43; English transl.: pp. 28–69, esp. 28). In his example, the ground is irreducible and continues to act under the form. The same can be said about the line in relation to the surface, from which the line marks itself out but which participates in defining the line. According to the author, difference is thus "an effectively real concept only to the extent that it designates catastrophes: either breaks of continuity in the series of resemblances or impassable fissures between the analogical structures" (ibid., p. 52; English transl.: p. 35). A difference shows itself being catastrophic also with respect to the numerical mode: the elements endowed with sense and designation are characterised by differential factors (cf. ibid.). Moreover, a difference is not equivalent to a mere opposition, because the latter introduces a reduction, a less profound relationship compressing complexity in a flat contradiction. This passage hence contains a critic to the dualistic articulation in reasoning.

---

[5]Cf. the chapter: "Aspect" in (Damisch 1995, pp. 77–90), in particular p. 81: "*en Occident comme en Chine un simple trait, tracé sur un support quelconque, peut faire sens, dans sa seule différence, son seul départ d'avec la surface (ce qui correspond bien à quelque chose, encore, comme un geste)*".

[6]Damisch refers to various writings of Ludwig Wittgenstein and to Jean Paul Sartre's *L'imaginaire*. He quotes in particular Sartre's definition of the image as the modality by which an object appears to the consciousness, that is, to put it another way, the modality by which the consciousness provides to itself an object (cf. Damisch (1995, p. 83), quoting from Sartre (1940, p. 17)): "The word 'image' could only indicate therefore the relation of consciousness to the object; in other words, it is a certain way in which the object appears to consciousness, or, if one prefers, a certain way in which consciousness presents to itself an object." Sartre (1940, English transl.: p. 7).

> It is not difference which presupposes opposition but opposition which presupposes difference, and far from resolving difference by tracing it back to a foundation, opposition betrays and distorts it. Our claim is not only that difference in itself is not 'already' contradiction, but that it cannot be reduced or traced back to contradiction, since the latter is not more but less profound than difference. On what condition is difference traced or projected on to a flat space? Precisely when it has been forced into a previously established identity, when it has been placed on the slope of the identical which makes it reflect or desire identity, and necessarily takes it where identity wants it to go – namely, into the negative. (Ibid., p. 73; English transl.: p. 51)[7]

Complexity requires thinking of differentials in multiplicity, avoiding dissipating the richness of the phenomenon through an understanding oriented by a pre-established synthesis. This is obviously the more difficult point to integrate in any theoretical and methodical approach to art, since it undermines the actual findings of an analysis by revealing the ideological viewpoint adopted. Or at least it discloses the underlying teleological expectations that any approach necessarily introduces to generate outcomes. To accept complexity, as an inescapable field with which the semiotics of art has to deal, means not only accepting the relativisation of one's results, insofar as they are dependent on the cultural framework that produces them. It also means accepting a dynamical approach that does not necessary aim to stabilise and close the interpretation on the basis of the aspects that the observation has enhanced. Concretely, what changes with respect to the, let's say, classic approach of the French semiotic school is that objects and phenomena—and especially art and aesthetic experience—cannot be questioned according to axiologies organised in oppositions, nor is it sufficient to conceive tensions between poles and dichotomies of values. Instead, it is more important to grasp the complexity of the phenomenon, analysable but not predictable, in which the emergence of meaning depends on a network of differentials.

If one really wants to consider a structural constraint in apprehending a work of art in its ensemble, one has to face that the gaze cannot separate but artificially the components of the perceived phenomenon. A work of art is an altogether of forms, colour, matter, light, spatiotemporal configuration, etc., and all these factors are generally intertwined in it. As Jean Petitot has suggested, it is a question of mereology[8]: parts are not only subjected to composition, but also interdependent in their own definition as specific components pertaining to the whole. The intertwining of plastic elements is an evidence as regard, for instance, matter, colour and light: colours depend on both matter and light; brightness as well as darkness are the result of colour interaction with the material support and the space-time positioning; matter in itself relies on colour and light which display or hide material properties,

[7] Deleuze writes here with reference to Hegel's philosophy, but contrasting as well with Trubetzkoy's declaration that the idea of difference presupposes the idea of opposition. He refers to Leibniz and Althusser to delve into his thoughts about differentials in a multiplicity.

[8] I owe this suggestion to the discussion with Jean Petitot who has raised the question in the debate following my lecture "Du chaos à l'(in)intelligibilité catastrophique de l'art", in the international conference "Actualité de René Thom. La dimension morphologique dans les sciences d'aujourd'hui", organised at University Paris 7 Diderot, on 7 December 2018 in Paris.

such as row roughness, texture, consistency or physic state. Moreover, the same proprieties of, let's say, matter, colour or light are, respectively, defined in relation to the other matters, colours or lights pertaining to the whole and their perceived material, chromatic or luminous values. For example, perception of a chromatic tone is dependent on the neighbouring colours; a shade is related not only to general light but first and foremost locally, to the adjacent enlightened areas. Also other less evident factors are interdependent in perceptual definition, not to mention in meaning foundation, viz. extent, size, distance and beholder's positioning, space-time articulation, and type of sensory solicitation. In addition, the perception of each component, of the entire work of art and of all their relations may vary according to previous perceptual knowledge, cultural context and background, and collective and individual disposition. Therefore, in the aesthetic experience each morphological element, at the very same time that it stands out and emerges in observation, is not dissociable from the relations it continues to have with the rest. It is apprehended as part of the whole, interacts with the sensitivity and disposition of the beholder and, in the network of these relationships, acquires its meaning.

Whereas all elements are part of a network of codeterminations and each one is thus ascribed to the continuity of that web, the work of art proceeds by catastrophes, by perceptual and signifying leaps, by "things that fall, the drops, the unbalances [*des choses qui tombent, des chutes, des déséquilibres*]", as Gilles Deleuze states in his 1981 lectures on painting, because "to paint, in a way, has always meant to paint local disequilibria [*peindre, d'une certaine manière, ça a toujours été peindre des déséquilibres locaux*]" (Deleuze 1981, session 1, circa 8′–9′, own transcription and translation). Indeed any change of form, whether in shape, chromaticism, texture, material or other plastic means, is a catastrophe, as it determines a boundary of that form. René Thom thus illustrates how every discontinuity in phenomena is a catastrophe[9]: "The edge of this table, the place where wood becomes air is a surface of separation, a catastrophe domain" (Thom 1991, new edition: p. 28; English ed.: p. 23). Even the boundary of a cloud, unless the cloud fades continuously into fog, is a catastrophe (cf. ibid., new edition: p. 29; English transl.: p. 24). So, passing from a morphological configuration to another, no matter whether figurative or not, material, perspectival or else, the gaze perceives boundaries and catastrophic transformations. Furthermore, since works of art mainly play on dynamic tensions, asymmetries, subtle or explicit imperfections, the sense of unbalance increases, to the point of empathetically involving the perceiver in local disequilibria or unstable balances.

Curiously, Deleuze does not pursue this idea of local disequilibria to think about catastrophe in art. By referring in particular to Turner, Cézanne, Van Gogh, Paul Klee and Bacon, the philosopher is mostly concerned in observing "a catastrophe that would affect the act of painting in itself", namely "a deeper catastrophe" than those represented as subject of the picture (Deleuze 1981, session 1, circa 11′ and 12′45″–12′55″ respectively, own translation). In his lectures on painting, Deleuze apparently hesitates many times on the issue but, actually, throughout his argumen-

[9]As Thom (1991, new edition: p. 23; English transl.: p. 23) states, "a catastrophe is a phenomenological discontinuity [*il y a catastrophe dès qu'il y a discontinuité phénoménologique*]".

tation he conceives the catastrophe merely insofar as it affects the morphogenesis of the painting. Either as a condition prior to painting or as an act undoing what has already been painted, the catastrophe frees the canvas from what precedes and burdens it before the very commencement of the painting. According to him, the catastrophe would apparently manifest itself in the work of art only in some specific cases, for instance, when Francis Bacon's *diagram* makes visible the potential of the figurable, or when *forces* modify, deform or shape the figuration, particularly in Michelangelo's art (cf. Deleuze 1981, session 2, 37″–1′05″).

However, in *The Fold: Leibniz and the Baroque* (1988), while he still declares to be mainly interested in the "genetic element" of the fold, Deleuze illustrates a schema of Klee on the active line and focuses on the "inflection" in the curve. As he points out, the inflection should be understood as the "elastic point" where the "active, spontaneous line" drawn by Klee transforms itself in a "variable curve or fold" (Deleuze 1988, p. 20; English transl.: p. 15). Belonging to the very phenomenal manifestation, the inflection point enables a morphogenetic shift. In other words, it allows different perceptual dynamics to take form and simultaneously coexist in the picture. As for Deleuze, the rereading of Klee's writings consistently encourages us to reinstate such a genetic moment into morphology, in accordance with the Goethean morphologic orientation, also inspiring the artist. In this way semiotics is able to pursue research on morphodynamics and heterogeneity in art.

Keeping in mind the idea of "local disequilibria": this is in particular the suggestion from Deleuze that can be upheld and expanded on, to consider the complexity of art—and the singularity of the works of art—in the light of Paul Klee's plastic thinking. The artist's writings extensively contribute to an understanding of how a work of art—as a phenomenon of aesthetic experience—proceeds by catastrophes, by perceptual and signifying leaps, while bifurcations in meaning, always possible, calls into question the intelligibility of art. To look, to observe implies to set a point from where the undifferentiated is transformed and from where the paths of complexity open up.

This is not the place to summarise Klee's theory, which will be analysed in the next chapter and which does not concern solely his art. Klee sets up a very comprehensive empirical method to analyse elements and consequent perceptual effects in works of art, based principally on two-dimensional arts, but which in many respects can be extended to art in general. With regard to the sensations of *weight* generated by plastic elements according to their quantities, qualities and relationships, Klee sketches scales to weight and balance shapes, colours, sizes … He also designs scales with multiple plates, spatial scales, even with different levels and heights, including highly unstable scales, such as the inverted pendulum with the equilibrium point at the top. His sketches can be compared to the diagrams of morphological dynamics with attractors and repellers. Advocating a balance that is such by including movement, i.e. also perceptual dynamics, Klee recommends transforming the scale into a spinning top, thus dynamically recording the differentials that, depicting form, chromaticism, texture, figuration, orientation, spatial arrangement, structuration, etc., compose the heterogeneity of the work. These dynamics are inscribed in the morphology of the work of art and configured within the relation with the beholder's perception.

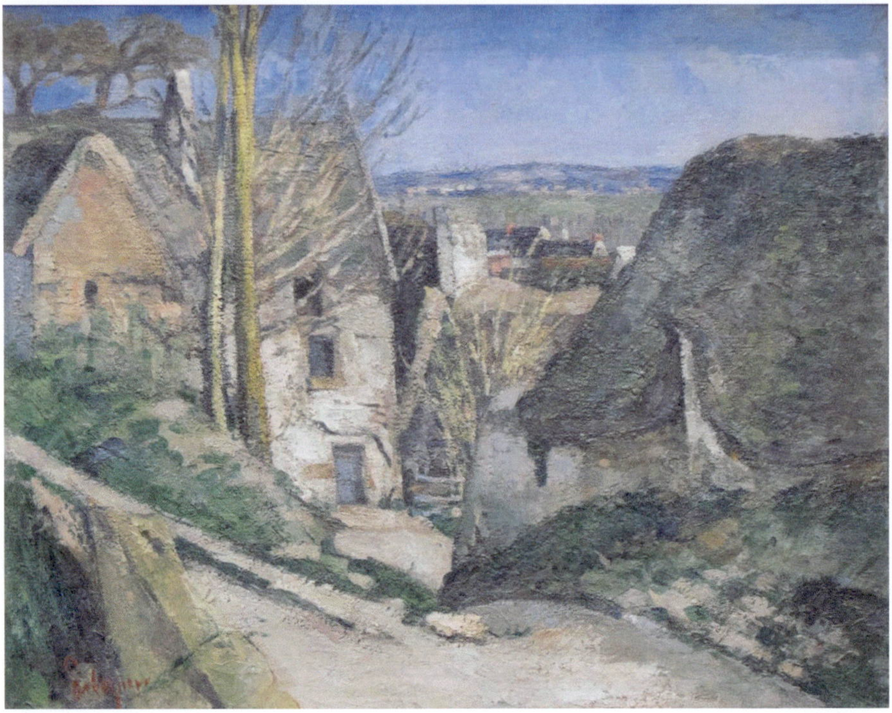

Paul Cézanne. *La Maison du pendu, Auvers-sur-Oise—The Hanged Man's House. Auvers-sur-Oise*. 1873. Oil on canvas. 55.5 × 66.3 cm. Musée d'Orsay, Paris

A look at one of Cézanne's paintings allows us to grasp what has just been discussed above. Any perceptual delimitation of form, colour, material, any variation in spatial and temporal morphology implies a catastrophe in the perceiver's observation. Works of art, whatever the artistic medium (painting, sculpture, installation, video, etc.), play on catastrophes, on local disequilibria, from which emerge lines, entities, coloured areas, plastic and figurative concretions, elements of composition, perspectives, spatiotemporal unfoldings.[10] In this case, the brushstrokes on the canvas configure forms, coloured shapes, sometimes delimited, sometimes blurred in their contours, often iconically recognisable, such as houses, trees, roofs, a village landscape relatively rural, arranged on several planes, from the path in the foreground to the hill backdrop, under a sky neither clear nor cloudy but warm. The tones, mainly in the ochre range, contribute to the overall smooth sensation of a balmy light, except perhaps the greens, spread over several elements and cooling down the picture, from the foreground and the roofs to the back field and even the sky. Without further describing the painting of Cézanne who, as Maurice Merleau-Ponty recalls, according to a certain criticism, "was abandoning himself to the chaos of

[10]The idea of space-time unfoldings in perception is taken on from Maldiney's philosophy, according to which a work of art creates, opens space and time to the beholder who perceives it.

sensations" (Merleau-Ponty 1945, new edition: p. 22; English transl.: p. 13),[11] let's note how the abrupt steep slope in the foreground creates a catastrophic leap in the path leading to *The Hanged Man's House* (*La maison du pendu, Auvers-sur-Oise*, 1873), pointed out in the title. An in-depth analysis can highlight the plastic and figurative traits that cause such a puzzling feeling in perception, thereby orienting to the main content of the painting. Various plastic elements draw the perceiver's attention towards the fulcrum of this leap. The profile lines of the principal depicted buildings theoretically converge towards the same point wherein the boundaries of the path lose their straightness and fold into the slope. The green strokes along the path, more spaced and blurred towards the border of the canvas, accelerate their rhythm in the direction of that point, thus attracting the perceiver's eye to it. Also the branches of the tree in the second plane, whose trunk is hidden by the house on the right, seem to pop up, gush out from this unfathomable point. Still, the barrier on the left corner, on which the painter has put his signature, delays, protects somehow from moving into the unsettling, while leaving the path open. But importantly, more than the plastic and figurative elements singled out, the contrasting perspectives give the full force of attraction to this blind spot, ultimately invisible, by a radical shift between the very foreground, the second plane and the background. The catastrophic scenario indicated in the title is thus depicted by the pictorial means themselves in their figurative as well as nonfigurative modes of expression. Clearly, the significance of this painting cannot be reduced to this sole issue, stylistic and other expressive features may be enhanced by historical and aesthetical interpretations. The present analysis aimed solely to emphasise a specific aspect concerning morphodynamics as related to art. Singularly, the local disequilibria not only articulate the emergence of forms in perception, but also combine each other to convey a semantic value in the representation.

Significantly, heterogeneity does not or not only result from the elements that compose the work of art. In this sense, even a monochrome painting does deploy a swarm a values. On the one hand, minute variations in texture, treatment and spatial positioning in the visual field relate to perception and, for instance, can generate different optical depths in retinal reception; on the other, any work of art takes on a form and acquires meaning with respect to a historical and geographical environment, thus meeting or reacting to certain expectations within the culture and the imaginary of its audience. Beyond analytical morphology, it is rather the set of differential and aspectual codeterminations that guide both the aesthesis and the significance of the work of art. While knowledge may orientate to a certain disposition in perception, perception integrates knowledge, activates it and even reconfigures it. To this end, from the multitude of differentials, *saliences* emerge, giving rise to forms, configurations,

[11]The philosopher mentions, then criticises the opinion of the painter Émile Bernard concerning the art of Cézanne, particularly between 1870 and 1890.

topological relationships, subject to structuring when one seeks their intelligibility—their *logos*, as Thom writes—and contributes thereby to establishing the meaning of art.[12]

> Finally I've reached the conclusion that, at the fundamental level, our minds must confront the world structure which, at its origins, consisted of salient entities. These entities emitted pregnances which other salient entities were able to receive or capture. In that context, effects were produced that one may call figurative, which might have induced the invested entity to re-emit the pregnance, or, depending on the situation, one that was slightly different. In reality, what we've got here is the ultimate source for all cases of systems in interaction. (Thom 1991, new edition: p. 109; English transl.: pp. 146–147)

Through the emergence of saliences and the reception or capture of pregnances, forms arise in perception and are defined within the interaction, involving the beholder, in increasingly developed levels ranging from psychophysical comprehension to clearly assumed interpretation. However, the evolution at every step is neither linear nor unidirectional, rather producing potentialities from what has been called the flow of heterogenesis. Indeed this idea allows a comprehensive understanding of the 'aspect' theory from a phenomenological viewpoint, and of polysemy in semiotics.

In a groundbreaking article titled "Differential heterogenesis and the emergence of semiotic function", the authors Alessandro Sarti, Giovanna Citti and David Piotrowski develop Deleuze's concept of differential heterogenesis and, with a mathematical anchorage, argue in favour of a morphodynamic theory explaining the emergence of semiotics. They observe how the construction of the semiotic function results from the dynamic evolution of the heterogenetic flow. Contrary to classic structuralism, in their opinion, a flow provides a consistent morphologic field without the need of any stabilisation, its flux "tends to create coherent forms, but these are continuously changing and are never stabilized" (Sarti et.al 2019, pp. 30–32, esp. 31). Pushing forward the Deleuzian complex theory of becoming, the authors conceive the rising of consistent configurations, that never stabilise, as the integration of a differential problem. In this regard, they propose the remarkable concept of vibration[13]: according to the modes of vibration proper to concatenation, also known as eigenvectors, the main axes of the flow are outlined and, thereby, the ontogenesis of the morphology and the semiotic configuration are defined. In a way, vibration seems to lead to the core of the co-constitution of meaning and sensitivity, and maintains open the fair play of interactions through which a meaning is attributed to the sensitive.

Congruously, Palo Alto scholars, already at the end of sixties (Watzlawick et al. 1967), highlighted how sensorial and cerebral activities are possible only through movement and the discernment of differences, that is, through the creation of relationships and patterns of relationships that found the experience. Whether in vision,

[12]In accordance with the customary terminology in the field and despite the lack of evidence, the terms 'salience' and 'pregnance' (or 'saliency' and 'pregnancy') are employed as a lean-translation of René Thom's French concepts of "*saillance*" (perceptual salience) and "*prégnance*" (biological relevance).

[13]The concept of vibration is of great significance in aesthetics and is being re-evaluated, particularly in relation to contemporary art. For an analysis of its theorisation in art, cf. Caliandro (2018).

hearing, touch, thermal feeling or any other sensorial form, the aesthesic comprehension results from the detection of relationships between discontinuities, which make similarities, dissimilarities and variations intelligible to senses and to brain. So, visual, sound, thermal … information can be grasped and elaborated only when sensory forms are comprehended into dynamics. "In other words, a relationship is established, tested over as wide a range as a given contingency allows, and an abstraction is eventually gained that, we hold, is identical to the mathematical concept of function. Thus, not «things» but functions are the essence of our perception" (ibid., p. 9 in 2011 edition; French transl.: p. 22).

The outline of an intrinsic function, namely a logos, a plastic and aesthetic logic, transforms percepts, articulated and concatenated each other, into signs of a possible significance, which is therefore both motivated and teleologically ascribed. To that extent, the vibration of perceptual flows resonates within the beholder through a response which is eminently empathic. In the following essays, the morphodynamics of the semiotic phenomena will be explored in relation to psychophysical values. Sensations and possible contents of works of art are examined as anchored to plastic means that convey them. Also the concept of empathy is reconsidered according to its historical aesthetic origins and thus developed in reference to art. Some analyses of works of art characterise the reflection in art theory and contribute to the general survey of the semiotic emergence of sense in art. Theoretical issues alternate with specific research on artistic production or on aesthetic concerns, e.g. focusing on spatiality, colour and the relation between feeling and saying. Beyond the theme addressed in each essay, an overall statement can herein be made to stress the singularity of the work of art. In this respect, the art phenomenon is not only conceived as singular in a philosophical sense. Moreover, it is not a matter of rehabilitating the secular understanding of the uniqueness of art since, as contemporary productions—and not only—have accustomed us to notice, criteria such as originality or non-repeatability are often not relevant to the aesthetic object, the latter being liable to be reproduced in series, created collectively, made in a participatory manner or even not to find any stable concreteness or physical persistence. The concept of singularity should then be intended in a morphological way: perceptual and signifying dynamics coexist in the aesthetic phenomenon and even conflicting dimensions can be found in the work of art. The logos, intuitively or semiotically defined, is not a structural element but rather proceeds to the structuring of forms through interaction with the sensory and cognitive comprehension of the beholder. Inasmuch as artistic reception opens up and facilitates the formation of various perceptual logics and modes of knowledge, the work of art—conceived as an aesthetic experience—contains different dynamics that cannot be reduced to a single understanding. As a figurative approximation to mathematics, a work of art can be said to constitute a singularity in that it concentrates multiple dynamics, is the elastic point in which the logics underlying these dynamics degenerate and inflect one another, and in which also the semiotic intelligibility folds into the sensitive.

References

Caliandro, Stefania: La vibration au tournant du XXe siècle et l'émergence de cette notion dans l'art. Paper in international conference: "Les cordes vibrantes" de l'art. La relation esthétique comme résonance, Fondation Singer-Polignac, Paris, 14 June 2018. https://www.singer-polignac.org/fr/missions/lettres-et-arts/colloques-arts-lettres/1361-les-cordes-vibrantes-de-l-art-la-relation-esthetique-comme-resonance#pcaliandro and https://vimeo.com/276190961, duration 28′ 34″ (2018)

Caliandro, Stefania: Sémiose esthétique: structuration et logos de l'art. L'anti-sculpture de Fausto Melotti. Paper presented in 2017 at the Congress of the Association Française de Sémiotique, UNESCO, Paris, published online in: Greimas aujourd'hui : l'avenir de la structure. Actes du congès de l'Association Française de Sémiotique, AFS Éditions, pp. 777–781. http://afsemio.fr/wp-content/uploads/AFS_Actes.2017.pdf (2019)

Damisch, Hubert: Traité du trait. Tractatus tractus. Réunion des Musées Nationaux, Paris (1995)

Deleuze, Gilles: Différence et répétition. Presses Universitaires de France, Paris (1968) [English transl.: Difference and Repetition. Columbia University Press, New York (1994)]

Deleuze, Gilles: La peinture et la question des concepts (session 1). Sound record of the lecture held on 31 March 1981, Paris 8. https://www.youtube.com/watch?v=vfRClbNdIFs, duration 1^{h}38′02″ (1981)

Deleuze, Gilles: La peinture et la question des concepts (session 2). Sound record of the lecture held on 7 April 1981, Paris 8. https://www.youtube.com/watch?v=4VOjdZiehI8, duration 2^{h}28′45″ (1981)

Deleuze, Gilles: Les plis de l'âme. In: Le pli. Leibniz et le baroque, pp. 20–37. Les Éditions de Minuit, Paris (1988) [English transl.: The folds in the soul. In: The Fold: Leibniz and the Baroque, pp. 15–29. Continuum, London, New York (2006)]

Édeline, Francis: Sémiotique de la ligne. Stud. Commun. Sci., 8, 1, pp. 189–213, Lugano (2008)

Groupe μ: Principia Semiotica. Aux sources du sens. Les Impressions Nouvelles, Brussels (2015)

Merleau-Ponty, Maurice: Le doute de Cézanne (1945). New edition in: Sens et non-sens, pp. 15–44. Les Éditions Nagel, Paris (1966) [English transl.: Cézanne's doubt. In: Sense and Non-Sense, pp. 9–25. Northwestern University Press, Evanston, Illinois (1964)]

Sarti, Alessandro; Citti, Giovana; Piotrowski, David: Differential heterogenesis and the emergence of semiotic function. Semiotica. J. Int. Assoc. Semiot. Stud./Rev. Assoc. Int. Sémiot., 230, pp. 1–34, Toronto (2019). https://www.degruyter.com/downloadpdf/j/semi.2019.2019.issue-230/sem-2018-0109/sem-2018-0109.pdf

Sartre, Jean-Paul: L'imaginaire. Psychologie phénoménologique de l'imagination. Gallimard, Paris (1940) [English transl.: The Imaginary. A Phenomenological Psychology of the Imagination. Routledge, New York, London (2004)]

Schapiro, Meyer: Paper presented in 1966. Semiotica, I, 3, pp. 223–242 (1969) [French transl.: Sur quelques problèmes de sémiotique de l'art visuel: champ et véhicule dans les signes iconiques. Critique, 315–316, Aug–Sept 1973, pp. 843–866. Reed. in: Style, artiste et société, pp. 7–34. Gallimard, Paris (1982)]

Thom, René: Prédire n'est pas expliquer. Entretiens avec Émile Noël. Éditions Eshel, Paris (1991). New revised and corrected ed.: Flammarion, Paris (1993) [English transl.: Tsatsanis, S.P. (ed.). Interviews with Emile Noël. Translation from the French, of Prédire N'est Pas Expliquer (To Predict is Not to Explain). Conversations on Mathematics, Science, Catastrophe Theory, Semiophysics, and Natural Philosophy. s.l. s.n. https://www.fermentmagazine.org/Stories/Thom/Predire.pdf (2010)]

Watzlawick, Paul; Beavin Bavelas, Janet; Jackson, Don D.: Pragmatics of Human Communication. A Study of Interactional Patterns, Pathologies and Paradoxes. W. W. Norton & Company, Inc., New York and London (1967). Reed. 2011 [French transl.: Une logique de la communication. Éditions du Seuil, Paris (1972)]

Wildgen, Wolfgang: Une entreprise audacieuse: Appliquer la théorie des catastrophes en linguistique/sémiotique. Paper in the seminar: "Morphologies and Qualitative Dynamics: An Interdisciplinary Project", Paris, 15 Dec 2017. https://www.actuality-renethom.com/, "Table ronde 02": "Linguistique, Sémiotique & Cognition" (2017)

Chapter 2
Paul Klee's Grey Point

Morphogenesis of Creation

In his artistic research and theoretical thoughts, Paul Klee was against any formalist approach that aims for a form bereft of function.[1] His aim was to render "the living form" (Klee 1956, English transl.: p. 60). For this purpose, the artist should be able to grasp and rediscover the creative act of the original genesis; he must be able to feel or, better, sense "the primordial living point"; as stated in his collected writings, "he knows of a little grey spot whence one can leap successfully from chaos to order" (ibid.).

In another short excerpt Klee enlightens on that "point in chaos: Once established[,] the grey point leaps into the realm of order" (ibid., p. 4), and he introduces in a few lines the pivot of his conception of art.

> Chaos as an antithesis is not complete and utter chaos, but a locally determined concept relating to the concept of the cosmos. Utter chaos can never be put on a scale, but will remain forever unweighable and unmeasurable. It can be Nothing or a dormant Something, death or birth, according to the dominance of will or lack of will, of willing or not-willing.
>
> The pictorial symbol for this 'non-concept' is the point that is really not a point, the mathematical point. (Ibid. p. 3)

This "locally determined concept", this mathematical "not-concept", is the catastrophic centre allowing the passage to the order from its opposite; it is the point where "the cosmogenetic moment" (ibid., p. 4) repeats itself, where the leap, or "the magic", "the mystery" (ibid., p. 60) of the art creation, occurs. It would be necessary to verify Klee's knowledge in regard to contemporary theories, to go deeper into the mathematical references he sometimes evokes in his writings, to see to what extent he was aware of the advancements of the Gestalt psychology, developed at the time

[1]This paper has been published in French in the Canadian academic journal *Visio* (Caliandro 2001–2002).

S. Caliandro, *Morphodynamics in Aesthetics*, Lecture Notes in Morphogenesis,
https://doi.org/10.1007/978-3-030-29631-5_2

in Germany and which seems to have prepared to a topological thinking[2] in the sense of the morphodynamism formulated later by the mathematician René Thom. Even while awaiting these historical confirmations, even giving a coincidence value to the interest that Klee, like Thom, showed in plant morphogenesis and organic biomorphism, it is rather hard not to be surprised by the correspondence of their research; to the point that, all other support failing, it easily comes to mind to declare the genius of the author, to notice the artistic intuition.

On many occasions Klee points out the limits of an analytic method,[3] he emphasises the "unweighable and unmeasurable" aspect of this catastrophic moment, of "a non-conceptual concept of freedom from opposition" (Klee 1956, English transl.: p. 3).

> If we express in terms of the perceptible (as though drawing up a balance sheet of chaos), we arrive at the concept of grey, at the fateful point between coming-into-being and passing-away: the grey point. (Ibid.)

The grey point is, for Klee, where the conflict between the opposites condenses. The grey point, static in itself, opens to contrasting factors and thus gives rise to the dynamism of the art creation. It would be possible to speak of different attractors that this primordial point brings into play. Catastrophic centre between the non-colours (black and white) as well as between the complementary ones (e.g. green and red), the grey becomes for Klee the very centre of the colour wheel, as shown in his spectral circle schema (cf. ibid., fig. p. 510):

[2]For instance, Wildgen (2001) supports the idea that Kurt Lewin's linguistic thoughts, imbued with Gestaltist theories, proceed from a non-quantitative use of new mathematical tools, that is, topology and vector analysis.

[3]In a pedagogic note for his lessons at the Bauhaus on 4 December 1923, he remarks upon the impossibility of using traditional mathematics to measure and compare movements, with a view to a proportional approach:

> We are not face to face with mathematics here. Nor is it a matter of fathoming measurabilities or weighabilities. We are concerned with comparing the impressions made by the various parts, and it is precisely the difference between at least two parts on which a higher proportion hinges. (Klee 1970, English transl.: p. 79)

Also in another note, he highlights the unavoidability of the impasses encountered by analytical reasoning:

> You have only to envisage something that is spatial to grasp the problems that necessarily stand in the way of an analytical approach. But how else is one to achieve orientation in space? I do not know!
>
> […] The real reason is that I am debarred from doing what I would rather do, discuss the whole at one stroke – this whole that embraces a very large number of these things, each in its place. Hence I have to settle for the analytical approach. (Ibid., p. 281)

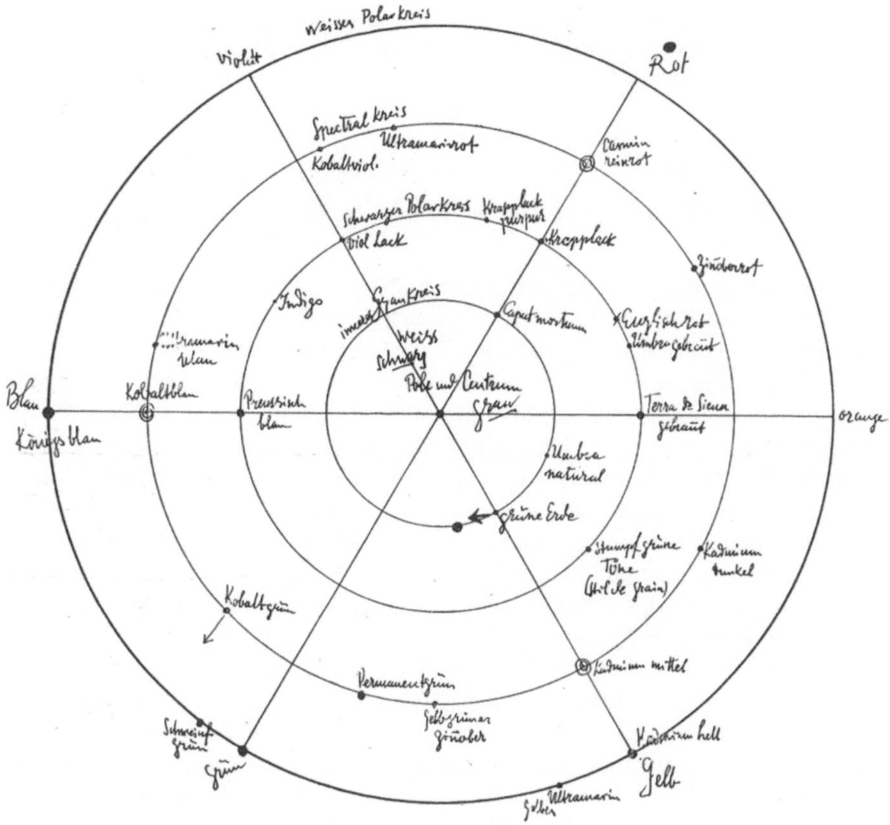

Paul Klee. Theory of pictorial configuration: 1.2 Principal Order. Pencil on paper. 21.6 × 27.5 cm. Zentrum Paul Klee, Bern, Inv. BGI.2/160. *Photo* © Zentrum Paul Klee

This point triggers hence conflicting relationships between various elements:

> The point is grey because it is neither white nor black or because it is white and black at the same time.
>
> The point is grey because it is neither up nor down or because it is both up and down. It is grey because it is neither hot nor cold; it is grey because it is a non-dimensional point, a point between the dimensions. (Ibid., p. 3)

It is not only in the middle of chromatic oppositions, hues and values, but also the principle of any topological and spatial deployment. Having no dimension in itself, it generates the energetic movements that lead from the point to the line, from the line to the surface and from the surface to the volume (ibid., pp. 18–24 and 103–113). It sets itself at the origin of any pictorial dimension while designating simultaneously the unstable equilibrium of these movements:

> The establishment of a point in chaos […], which, concentrated in principle, can only be grey, lends this point a concentric character of the primordial […]. The order thus created radiates from it in all directions […].

> When central importance is given to a point: this is the cosmogenetic moment. To this occurrence corresponds the idea of every sort of beginning (e.g. procreation) or better still, the concept of the egg. (Ibid., p. 4)

Unstable Equilibrium (or the Stability of Dynamism)

Klee takes up in many respects the idea of the scale that structures the complex articulation of tensions at work. The scale helps stabilise the overall organisation of the image, measure the dynamism that various elements engender in perception. Visual composition tool, it manifests and, whether suitable, rebalances the instabilities of an asymmetrical configuration. Sensitive to the difference in extent, colour and tonality of the various parts, it enables the counterbalance of these discrepancies in size, quality and visual weight to make them interact in a dynamic equilibrium (ibid., pp. 197–214). The structural stability that ensues is never for Klee the result of a static organisation; on the contrary, it implements contrasting tensions whose final equilibrium is continually called into question.

Many are the tightrope walkers, jugglers and equilibrists that populate the artistic world of Paul Klee, as in *Tightrope walker* (1923), *Height!* (1928), *Odd theatre* (1929) or *The devil, juggling* (1930). In *Head jugglers* (1916), the different fragmentation and recomposition of the bodies of the personages, their position off or on the edge of the line, their place on the surface of the image create echoes and correspondences which are often contradictory or reversed. For each personage, more or less important deviations are drawn in relation to the upright vectoriality of his figure. The set, however, composes a dynamic equilibrium, of which the circle at the top centre is the symbol. The personage at the right centre, slightly deflected towards the right, leaves also his head to slide there; the raised arm of the neighbour on the left and his head relocated on his hand on the left provide some equipoise. But the equilibrium becomes even more precarious and complex as soon as one moves towards the lateral sides of the composition. On the extreme right, a juggler with his head on his belly, the body perhaps half overturned, exhibits a plastic heaviness through reiterated lines, which contrasts with his figurative lightness on the rope. To this man, echoes the couple on the opposite left side where the fragility of the upper juggler, upside down and in difficult balance on a stick, contrasts with the decided stability of the personage sitting on the ground.

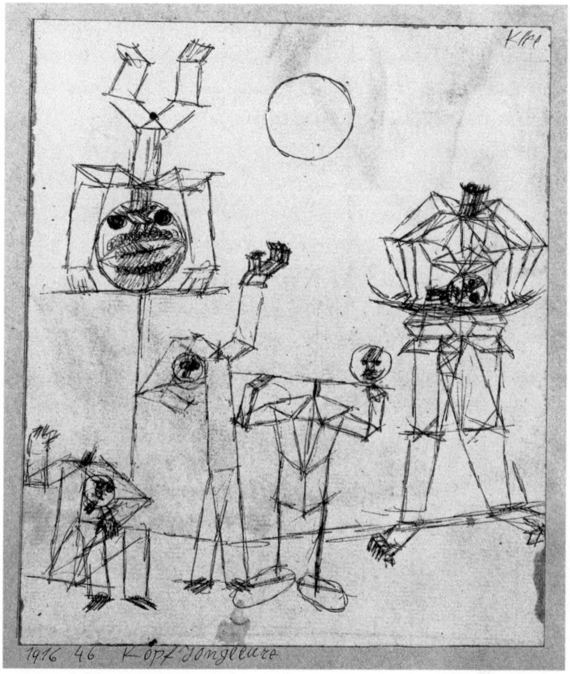

Paul Klee. *Kopf Jongleure—Head jugglers*. 1916, 46. Pen on paper on cardboard. 17 × 14 cm. *Photo* © Zentrum Paul Klee

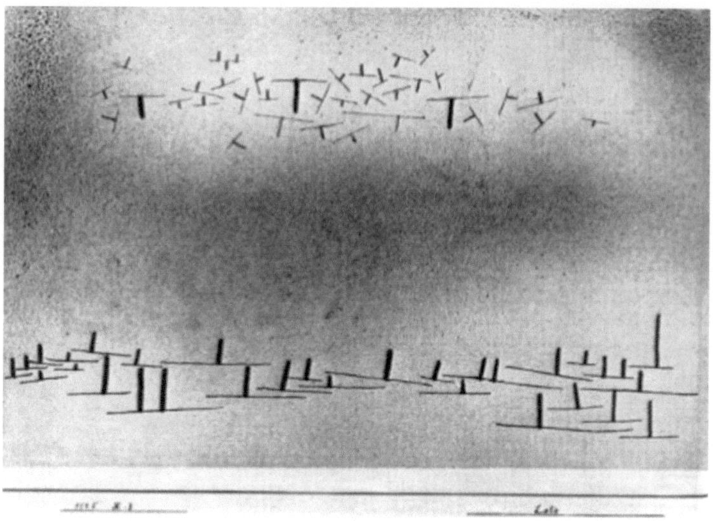

Paul Klee. *Lote—Plumblines*. 1925, 233. Pen and watercolour on paper on cardboard. 20.1 × 30.9 cm. Swiss private collection, deposited in Zentrum Paul Klee, Bern. *Photo* © Zentrum Paul Klee

At other times, it is only the tensions between colours and/or shapes and their spatial arrangement that determine a dynamic equilibrium, for example in *Daringly poised* (1930)[4] and *Le rouge et le noir* (1938). In *Plumblines* (or *Lote* in German, 1925), the different inclination of the perpendiculars at the top and bottom is sufficient to suggest a freer, more aerial floating in the upper area, whereas in the lower part, the oscillation looks more dependent on the gravitation laws and seems to allude, as some critics have indicated, to the rocking of the ships on the water. If, in this case, the composition, approximately symmetrical with respect to the vertical median, plays on the variations of size, location and rotation to suggest the weight and the consequent movement of the forms, elsewhere the symmetry is committed to fix the expressiveness of an anthropomorphic figure. In *Senecio (soon to be aged)* (1922), the inclination of the geometric elements that define his gaze contrasts with the more broadly static equilibrium of the image. However, the raised eye on the left is counterbalanced by the weight of the superposed triangular shape, whilst the eye on the right, deeper in the face, receives a new momentum thanks to the circular arc on the eyebrow.

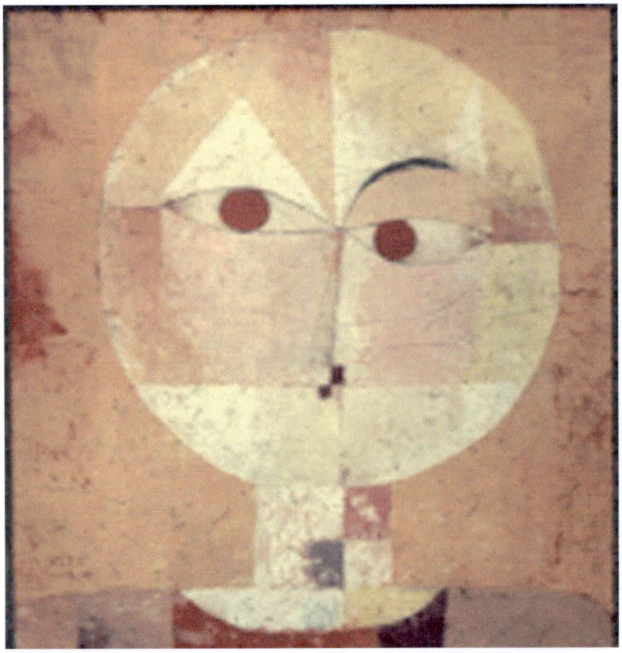

Paul Klee. *Senecio (Baldgreis)—Senecio (soon to be aged)*. 1922, 181. Oil on primed gauze on cardboard. 40.3 × 37.4 cm. Kunstmuseum Basel

Other paintings also work on these contrasts which give to forms the tension of an unstable equilibrium and, therefore, the impression of animated forms. Furthermore,

[4]Reproduction dans Klee (1956, English transl.: p. 213).

Klee—and this is one of the fundamental points of his research—will focus significantly on this threshold between abstraction and figuration, between the opacity of visual forms and the transparency of the representational sign. Not interested in pure abstraction, wary of any empty and sterile formalism, he defines himself in this way in his diaries: "I am «abstract with memories.»" (Klee 1957, English transl.: p. 315). In general, the allusion, the evocation of the world prevails in his creations. Even when he devotes himself to works of art devoid of any figuration—which is the least frequent case—he aims to depict "the living form" (Klee 1956, English transl.: p. 60) and to animate forms and colours through perceptive tensions. Lines and surfaces, thus, acquire a vital breath, they are personified. A real actualisation of the empathic projection appears to be at work. Klee seems to exploit the relationship between the contemplated object and the beholder. Aware of the book *Abstraktion und Einfühlung* by Worringer (1908), also mentioned in his diaries, Klee nevertheless seems to connect these two poles of abstraction and empathy that Worringer wanted to be opposite. In this sense, he was resuming the aesthetic theories in visual perception, which, since the end of the nineteenth century, have envisaged the animation of inanimate or even abstract forms through empathy.[5]

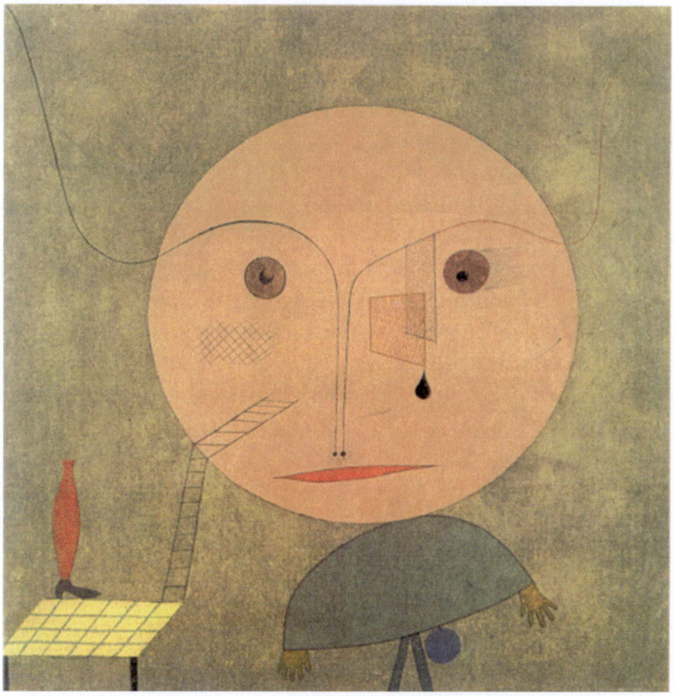

Paul Klee. *Irrung auf Grün—Error on green*. 1930, 182. Watercolour on cotton. 40 × 38 cm

[5]For an historical overview concerning the theories of empathy, cf. please Caliandro (1999).

Error on green (1930) produces precisely this tension between, on the one hand, simplification, geometrism, reduction to minimal visual elements and, on the other, a sketch of a figuration. The detached leg, the ladder rising to the face, the eyes in the shape of a black moon and possibly a black sun, the falling drop like a black tear amplifying the distress of those atrophied hands: the openings towards a symbolic reading are very present, and yet the latter is thwarted by the perfect circularity of the visage, by the schematism of the body, the absence of volumetry or depth, the extreme stylisation of the nose and eyebrows which, in a single line, exceed the limits of the face and hang from the picture's upper corners. But, upon closer examination, the tension between figurative and merely visual elements is nurtured by another tension, this time strictly nonfigurative, connecting the parts to each other. The size of the face, out of the barycentre, creates a figurative instability that makes the body of the personage lean towards the left and which, to no avail, the segmented ladder and the weak eyebrows try to rebalance through external supports. Only the decentring towards the right with respect to the format of the picture and, above all, the small circle on the right, next to the personage's legs, restore a perceptive counterpoise; all the more and since the blue of the circle provides a satisfactory visual counterweight to counterbalance the extent of the other surfaces.

Increasingly, multiplying the pairs of plates that must be weighed, the balance quickly turns into a multi-branched instrument, of "[f]ully spatial scales" (Klee 1956, English transl.: p. 199), whose resulting equilibrium becomes more and more complex. Charged for the stabilisation of highly articulated measurements, qualities and hues, "[n]o matter how accurately the weights are divided, the standing scale supported by the earth at one point only will totter and fall" (ibid., p. 385). The sole way to avoid the tumble is then to animate it with an energetic movement, to transform it in into a spinning top. "Every child knows – explains the painter at the Bauhaus – that a top does not fall as long as it remains in motion, as long as it spins" (ibid.). Thus, Klee introduces in some paintings a "visible dynamics" (ibid., p. 384) by increasing or diminishing in quality or quantity the energy of the visual elements. Vectorialities emerge and animate the surface in several directions, as in *Rosewind* (1922) where the continually solicited gaze of the beholder swirls according to the senses given by shapes and colours.

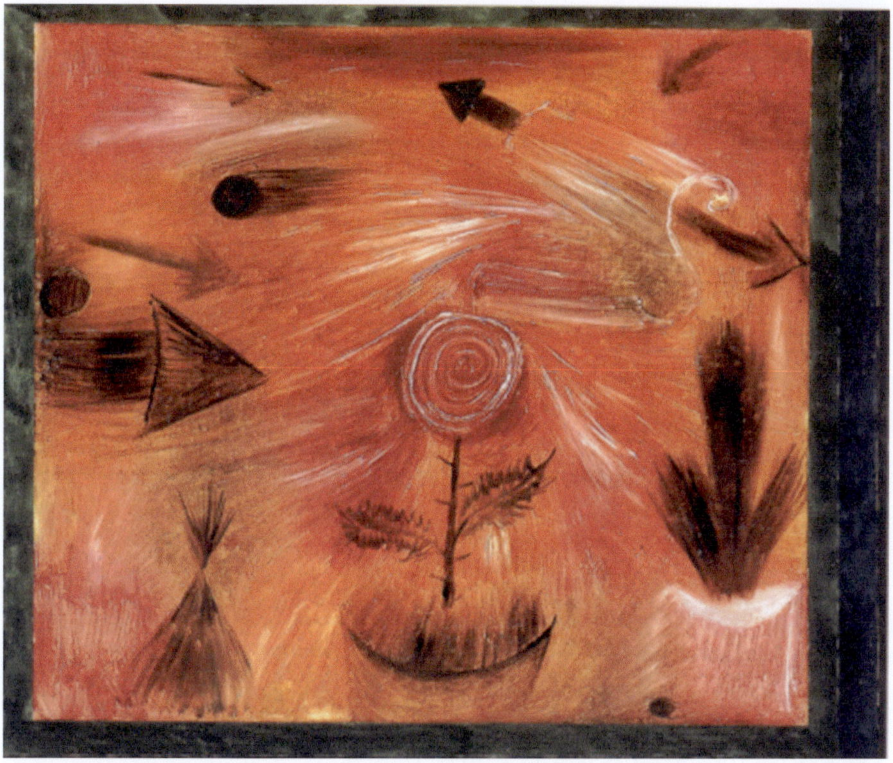

Paul Klee. *Rosenwind—Rosewind*. 1922, 39. Oil on primed paper cardboard. 38.2 × 41.8 cm. Zentrum Paul Klee, Bern, donation Livia Klee. *Photo* © Zentrum Paul Klee

Following these thoughts on the "visible dynamics" (ibid.) and in order to give rise to perceptual movements into the still image, Klee uses the idea of the pendulum.

> In between, between statics and dynamics, lies an intermediate realm, whose symbol, the pendulum, represents a compromise between the two realms. (Ibid., p. 387)

The pendulum indeed connects statics and dynamics, gravity and momentum, rest and movement. It is certainly not by chance that this same instrument will also become a key element to represent the morphodynamic evolution in topology. In the pendulum, according to Klee, the concept of oscillation is essential: it marks the combination of movement and its immediate countermovement. But the artist is equally sensitive to the difference between the pendulum whose ball of weight swings below and the inverted pendulum, i.e. a metronome,[6] which on the contrary has down

[6]Klee's passion for music is well known, making him hesitate for a long while between working as a painter and as a musician. Even after his decision, Klee pursued his passion, in particular by playing the violin.

its oscillatory centre. In morphodynamics, the first type of pendulum, "stable at the bottom and unstable at the top" tends towards a centre of rest, which constitutes its "attractor", while in the second case, the slightest perturbation makes the ball escape from the summit, which becomes the "repellor" (or repeller; Wildgen 1999, chap. 2, § 2).[7]

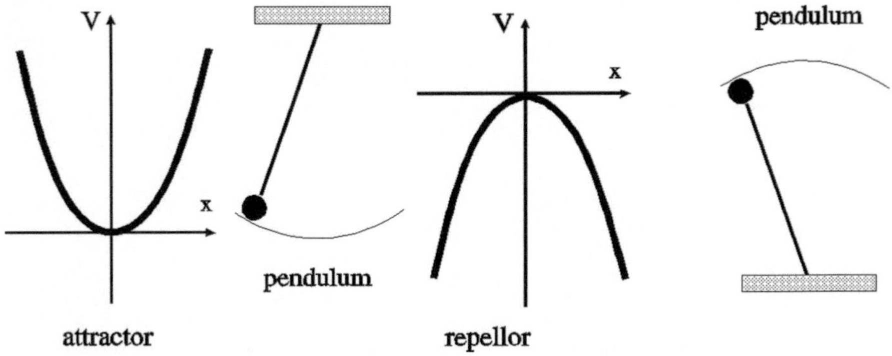

Wolfgang Wildgen. Basic dynamical systems. *Source* © Wildgen (1994), fig. 3.3, p. 50

Similarly, Klee had been led to recognise a "[s]tatic tension passive" in the rest position of the classic pendulum and to contrast it to the perfect verticality of the inverted pendulum where the "[s]tatic tension active" is the result of a fragile equilibrium (Klee 1956, English transl.: p. 393). Thus, the movements arising from these two different static positions transmit the sensation of either a fall or a construction (cf. ibid., fig. p. 393: static, static-dynamic and dynamic tensions).

[7]Own translation from the excerpts, sent by the author, from the prepublication final version of the book. Cf. also the text with the Fig. 1: Two basic types of dynamics: stability (attractor) and instability (repellor), available at: http://www.helsinki.fi/varieng/series/volumes/03/wildgen/.

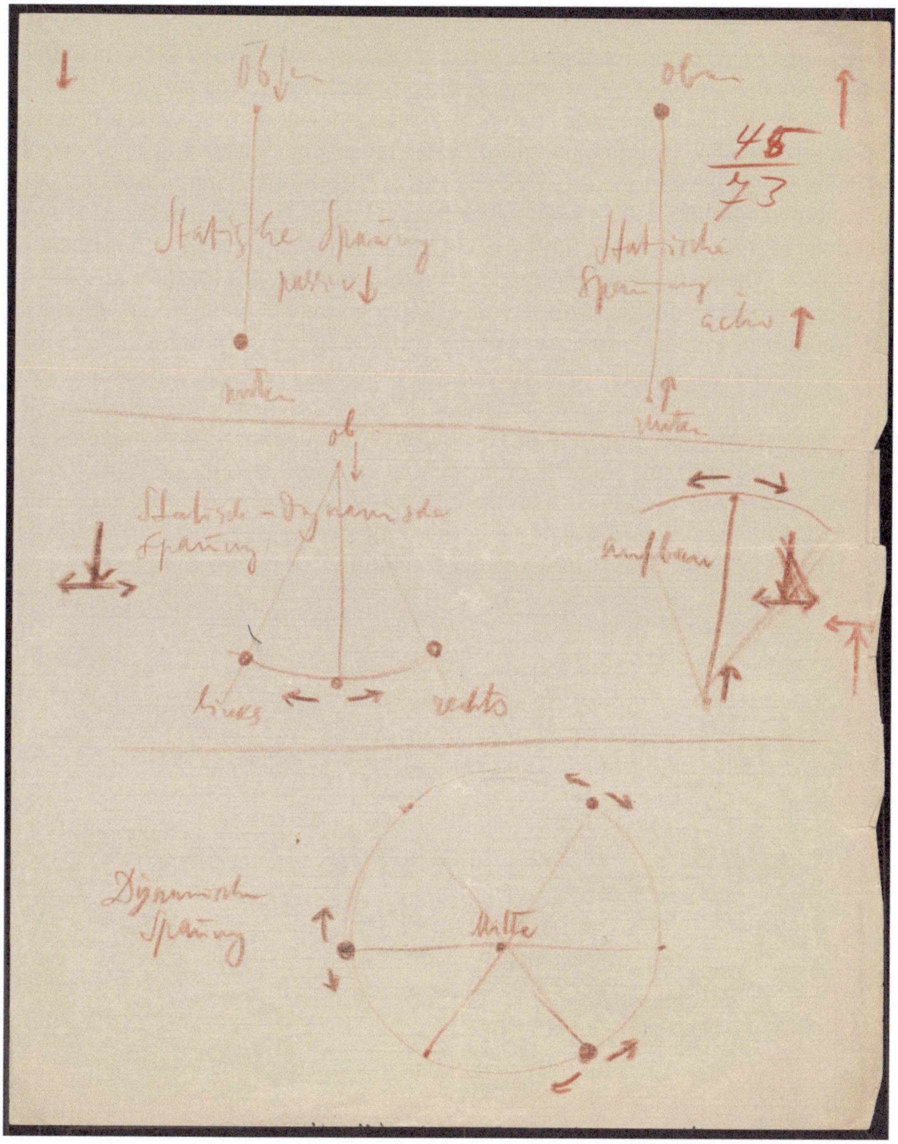

Paul Klee. Theory of pictorial configuration: II.21 Mechanics. Coloured pencil on paper. 28.9 × 22.2 cm. Zentrum Paul Klee, Bern, Inv. BG II.21/108. *Photo* © Zentrum Paul Klee

These thoughts reappear in his practical research in which the beholder—by way of *embodiment*, as cognitive sciences would say nowadays—is led to see in the pictorial forms the same rules that organise the physical world. The painting *Unstable Equilibrium* (1922) plays on the restoring of and the deviation from the law of terrestrial gravitation. The inclination of the different layers draws ascending and descending paths on which vectors are installed. These, like the two large arrows pointing up and down, slightly deflect from the orthogonal axis. The rectangular areas seem to counterbalance each other for form, colour, tonality and spatial arrangement on the surface. Thanks to their chromaticism (white, yellow, cyan), the transparency effect and the lack of filling of a few internal nets, they acquire an airy lightness, which contrasts with the chromatic materiality of the layers. All these different visual balances finally encourage the oscillation of the gaze following a back and forth of nonfigurative repetitions and variations. The perception retrieves therefore the same impressions that governed the pendulum's movements: the momentum, the resistance, the return, the fragility of the central equilibrium. This perceptual oscillation can finally be correlated with the circularity of the abovementioned picture, *Rosewind*, wherein the swirling of the gaze is rather comparable, in Klee's terms, to the "dynamic tension" (ibid.) of a circular pendulum free to turn completely around its anchor point.

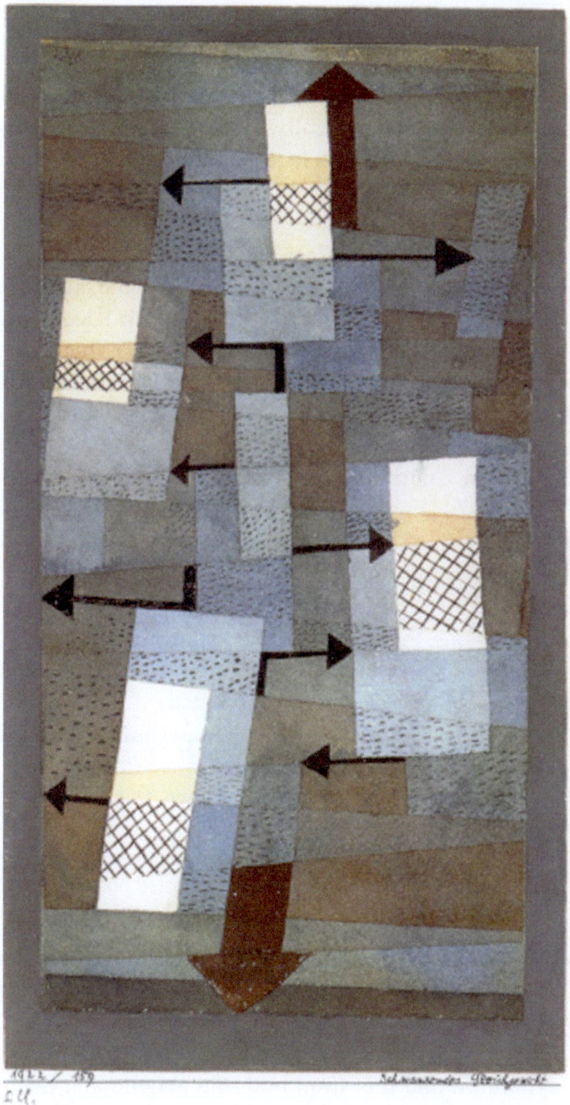

Paul Klee. *Schwankendes Gleigewicht—Unstable Equilibrium*. 1922, 159. Watercolour and pencil on paper on cardboard. 31.4 × 15.7/15.2 cm. Zentrum Paul Klee, Bern. *Photo* © Zentrum Paul Klee

Morphodynamism and Polyphonic Painting

This attention to the perceptual movements in the work of art stems from a morphogenetic conception that more generally links the different phases of the artistic creation. The same movement unites in fact what Klee calls "pre-creation, creation, [and] post-creation", namely: "The initial movement in ourselves; the active movement from us to the work; the communication of the work's mobility to others, the beholders of the work" (ibid., p. 169). Each successive phase keeps track, memory of the previous. The relation between formation and form is thus retained in later stages. This genetic trace, this idea of "memory" through which Klee defined himself—"I am «abstract with memories»", i.e. not an abstract painter, but simply "abstract", to mark the relationship that had with his life, with the war that took away his friend painter Franz Marc and from which he wanted to detach himself[8]—this idea seems quite equivalent to the "survival manoeuvre" of a catastrophic transformation that, according to René Thom, "permits the system to subsist when it would normally cease to exist" (Thom 1974, English transl.: pp. 89–90). To detach himself from his problems, the artist recognised himself in his work; up to identify himself with his own style, even when the latter was not yet shaped, as evidenced in his diaries on 22 June 1902:

> I am starting to learn all over again. I begin to execute forms as if I knew nothing about painting. For I have discovered a very small, undisputed, personal possession: a particular sort of three-dimensional representation on the flat surface.
>
> And when the night comes I can lie down with the consciousness that the work has been accomplished. And that also means something.
>
> A flying man! I force the third dimension into the flat plane. Disposition of the arms, paired legs, absence of foreshortening.
>
> I even dream about it. I dream of myself. I dream that I become my model. Projected self. Upon awakening, I realize the truth of it. I lie in a complicated position, but flat, attached to the linen surface. I am my style. (Klee 1957, English transl.: p. 124)

In the same way, that the author is led to project himself into the impending form, so art keeps in itself the work of formation that makes it come alive and that the beholder rediscovers each time in a different manner. The formation then predominates over the form understood as a result, accomplishment. What matters is the becoming of the forms, their dynamics, and this from both the genetic and perceptual points of view:

> The relation between formation and form, which we discern even in the smallest things, retains its fundamental character in later stages, because it is determined by a principle.

---

[8]Let me quote extensively this already mentioned 1915 excerpt from Paul Klee's diaries:

> I have long had this war inside me. This is why, interiorly, it means nothing to me.
>
> And to work my way out of my ruins, I had to fly. And I flew. I remain in this ruined world only in memory, as one occasionally does in retrospect.
>
> Thus, I am "abstract with memories". (Klee 1957, English transl.: pp. 313–315)

> I think the nature of this relation can be stated in one sentence: the way to form, dictated no doubt by some inward or outward necessity, is higher than its own end and goal.
>
> [...] Thus form may never be regarded as solution, result, end, but should be regarded as genesis, growth, essence. (Klee 1956, English transl.: p. 169)[9]

From that proceeds Klee's interest in the phenomena of biological growth and vegetal evolution, such as plant morphogenesis which inspires many of his works (ibid., pp. 21–23).[10] But this leads him to explore also the representation of the movement in painting, as in *Fugue in red* (1921). The problem of movement in a still image concerned many artists at that time, from Marey's chronophotography to Futurists and Duchamp. Klee, however, considers it rather in much the same way as Delaunay, less through the figuration of spatially displacing forms than through the optical movement given by colour. He develops the comparison with music already outlined by Delaunay and Kupka, as well as Kandinsky. Instead of looking for correspondences between sounds and colours, timbres and tones, he delves mainly into the differences.

In the text "Creative Credo", published as an article in 1920, but continuously exploited and revised for his teaching at the Bauhaus, Klee denounces the "good deal of fuss" ensuing from the distinction, theorised in Gotthold Lessing's *Laocoon*, between temporal and spatial arts. According to the artist "space itself is a temporal concept" (Klee 1956, English transl.: pp. 76–80, esp. 76). "The pictorial work springs from movement, it is itself fixated movement, and it is grasped in movement (eye muscles)." (Ibid., p. 78). The only difference between music and painting resides in the order of the perception, pre-established by succession in the former, variable in the contemplation of the latter (cf. ibid.). Yet, in his personal thoughts, Klee had already gone further. In his diaries, on 10 July 1917, he writes:

> Simple motion strikes us as banal. The time must be eliminated. Yesterday and tomorrow are simultaneous. In music, polyphony helped to some extent to satisfy this need. [...] Mozart and Bach are more modern than the nineteenth century. If, in music, the time element could be overcome by a retrograde motion that would penetrate consciousness, then a Renaissance might still be thinkable. (Klee 1957, English transl.: p. 374)

Instead of figuring time in art, Klee seems mainly interested in the possibility that painting offers to render a movement devoid of any temporality. Indeed, it is not the temporal factor in the representation that intrigues him, but the movement or, more precisely, the dynamism that the form can generate in perception. In this sense, his emphasis on the formation of form, on the morphodynamic aspect of the objects,

---

[9]Few lines later pursue as below:

> Form as movement, as action is a good thing, active form is good. Form as rest, as end, is bad. Passive, finished form is bad. Formation is good. Form is bad; form is the end, death. Formation is movement, act. Formation is life. (Klee 1956, English transl.: p. 169)

[10]Not only did Klee manifest a deep passion for plants, which he cultivated in his herbarium, but in his diaries, he also wrote to "lavish pious attention" on a problem of structural understanding such as the inversion of the sap capillary flow in the bergamot cutting (Klee 1957, English transl.: p. 204).

will be better understood. Likewise, in view of this becoming of the form, of this dynamism of the work of art, all the strength of his famous phrase can be reassessed: "Art does not reproduce the visible but makes visible." (Klee 1956, p. 76). Not only is art not a naturalistic imitation but is also nor just visible: it "makes visible", precisely through this timeless movement that links formation to perception, to the deployment of the form in the beholder's (empathic, it may be specified) reception. In the same perspective, Klee formulates and then practises his "polyphonic painting" (e.g. *Polyphonic setting for white*, 1930; and *Polyphony*, 1932):

> Polyphonic painting is superior to music because its time is more spatial. The idea of simultaneity comes out more richly. The reflection in the side windows of a moving tram car gives an idea of the backward movement I have in mind for music. (Klee's 1917 note quoted ibid., p. 520)

These images in the side windows of the tram fascinate the artist not, apparently, for the speed of the vehicle distorting the elements of the landscape. It is the reflections that charm him, these flat and still reflections, probably from inside the tram, which are projected on the transparent surfaces, on the surrounding landscape that those 'windows on the world' allow to see. It would actually be a matter of conflicting perspectives, points of view, representations spatially contrasting. The windows let see various perceptions of the surrounding space; they enable the catastrophic leap—while leaving them coexisting—between different spatialities: between the image of the external landscape and the reflection of the interior, between the three-dimensional depth and the flatness of the specular surface, between the transparency of the window and the opacity of the material support. Klee has long worked on spatiality in painting and almost all his production seems to rethink the way to render spaces that are often conflicting between them. Already in 1902, when he identified himself with a style he was still far from having developed, he was considering "a particular sort of three-dimensional representation on the flat surface" (Klee 1957 English transl.: p. 124, quoted *supra*). Much later, he manages to figure it in more and more complex pictures. In *Agitated landscape with spherical trees (agitated, brightly coloured surfaces)* (1920), he paints a flat two-dimensional landscape where movement is just suggested by the rhythm of the trees and the extents. However, memories of classical perspective, might it be a bird's-eye view, are evoked locally, for instance, in the lower left corner or in the centre slightly to the left; here, the proportions of the trees allude to a depth, which contrasts with the flatness of the rest of the picture and generates a disproportion of some elements with the whole. But certainly *Room perspective with inhabitants* (1921) shows more clearly the simultaneity of different spatial representations. The picture calls into question the principles of the linear perspective, not only because of the multitude of lines revealing and denying the artifice of the perspectivist construction (the vanishing point moves over and over), but also for the paradoxical shifting of the space below the horizontal line, wherein the inhabitants are set, fragmented and flattened as they are in a two-dimensional projection.

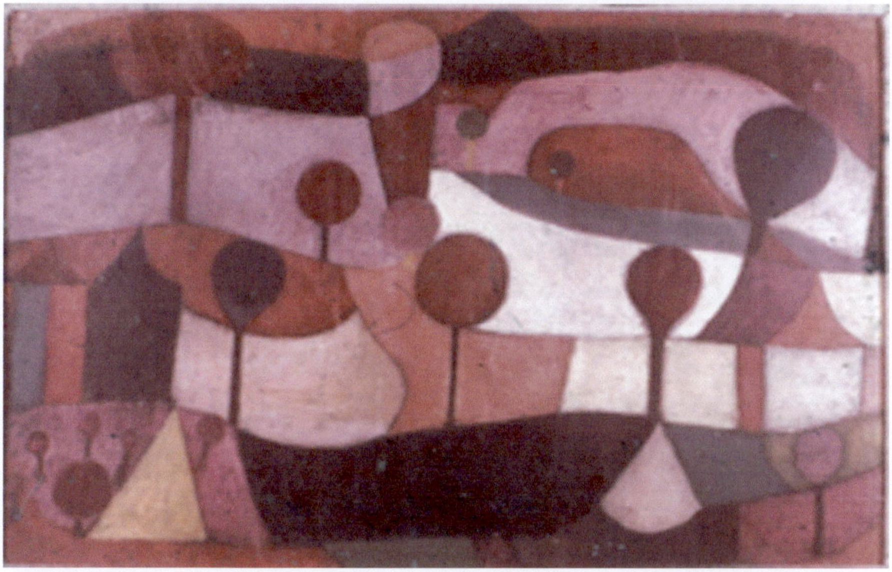

Paul Klee. *Bewegte Landschaft mit Kugelbäumen (bewegtestark farbige Flächen)—Agitated landscape with spherical trees (agitated, brightly coloured surfaces)*. 1920, 124. Oil on cardboard. 31.5 × 39.5 cm. *Photo* © Zentrum Paul Klee

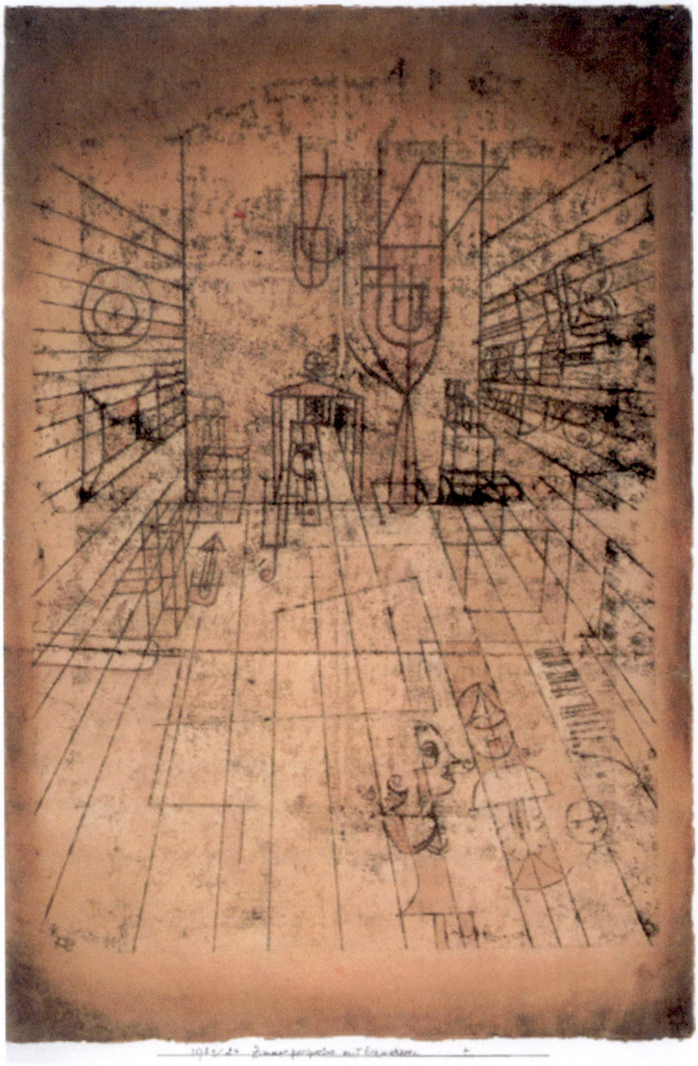

Paul Klee. *Zimmerperspective mit Einwohnern—Room perspective with inhabitants*. 1921, 24. Oil transfer drawing and watercolour on paper on cardboard. 48.5 × 31.7 cm. Zentrum Paul Klee, Bern. *Photo* © Zentrum Paul Klee

The composer and director of orchestra Pierre Boulez, definitely one of the commentators most attentive to these perceptual tensions, authored a book in which he non-technically and ardently links Klee's two main passions, painting and music. Rereading thoroughly his pedagogic writings, Boulez analyses the overturns made by the painter to a mode of vision that had dominated since the Renaissance:

> Starting from an example that could not be simpler: the railways, Klee proposes to his pupils to figure out the play of multiple perspectives, perspectives moving according to the point of

> the picture where an imaginary beholder would be. If the latter is on the right, the depicted perspective will not be the same as if he were on the left or centre. That is the side optics. But there is also a vertical optics. Klee combines various operations by means of horizontal planes at different levels and vertical planes variously placed from right to left. We quickly arrive at a mobile space with three dimensions. He shows his pupils how to construct a space whose centre moves, and the effects that this displacement may have on colour intensity. He emphasises the importance of the eye level, the consequences for changing the relation between height and horizon.
>
> Klee has constructed several pictures based on these principles, using multiplied perspectives, spaces succeeding one another, interpenetrating each other. (Boulez 1989, pp. 55–64, own translation)

Nonfigurative elements initiate perceptive tensions between several pictorial modes given simultaneously.[11] In *Architecture (yellow-violet stepped cubes)* (1923), the chromatic chessboard arithmetically arranged in a flat disposition acquires a problematic and undecided volumetry in the detail left down to the centre. Klee applies indeed the principle that he counselled his pupils "to create deviations while bearing the rules in mind" (Klee 1956, English transl.: p. 152). It could be said that choosing a predominant pictorial mode, here the geometry regularly organised on the surface, the artist creates a background on which perceptual saliences stand out problematizing the whole. To escape the barrenness of a uniformly applied law, he reworks the chosen pictorial mode by questioning it from inside, at the very moment he sets it up. Like the cubic forms at the bottom but for other aspects, the two triangular tips at the top suffice to put the two-dimensional abstraction of the picture into question. Without challenging its structural geometry, they evoke an urban figuration that reconfigures the entire chessboard. Far from applying the theory of the pendulum only to hues, colours and extents, Klee lets it oscillate between further pictorial settings: between two- and three-dimensionality, between contrasting perspectives, between figuration and abstraction. Just as in morphogenesis every dynamics reveals to be "endowed with a memory" (Brandt 1992, p. 280, own translation),[12] Klee seems to fix in a work of art the dynamic position of the pendulum which, substantially set towards a pole, nevertheless retains the memory of the momentum or repulsion towards the opposite pole.

---

[11]See also the followings notes in which Klee synthesises how to deviate from a pure central perspective:

> Organic combination of the main forms of perspective:
>
> interpenetration of space and body, simultaneous inner and outer form. Representation according to essence and appearance.
>
> Points to be considered:
>
> Simultaneous, multi-dimensional phenomena.[...] Multi-dimensional contacts. More complex structures. (Klee 1956, English transl.: p. 155)

[12]As Brandt explains, a dynamic "is able to make itself a representation, an image, of the path g that is at the base of its existence" (Brandt 1992, p. 280). Brandt puts forward the hypothesis that when a dynamic meets other dynamics, "these too, provided with memory", they can exchange each other their images as well as the values of their nervousness (ibid., p. 281).

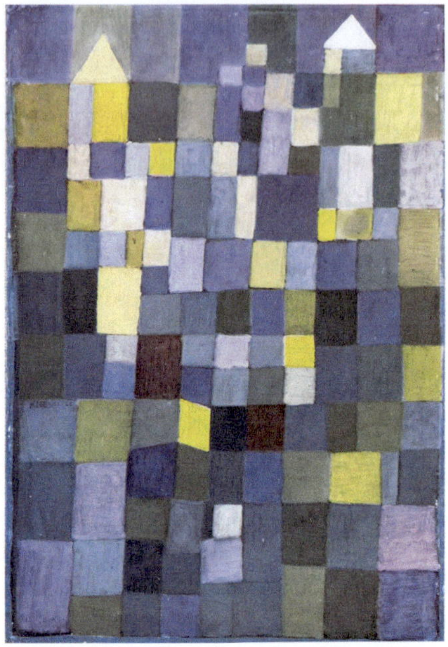

Paul Klee. *Architektur—Architecture (yellow-violet stepped cubes)*. 1923. Oil on primed cardboard. 58 × 39 cm. Nationalgalerie, Staatliche Museen, Berlin

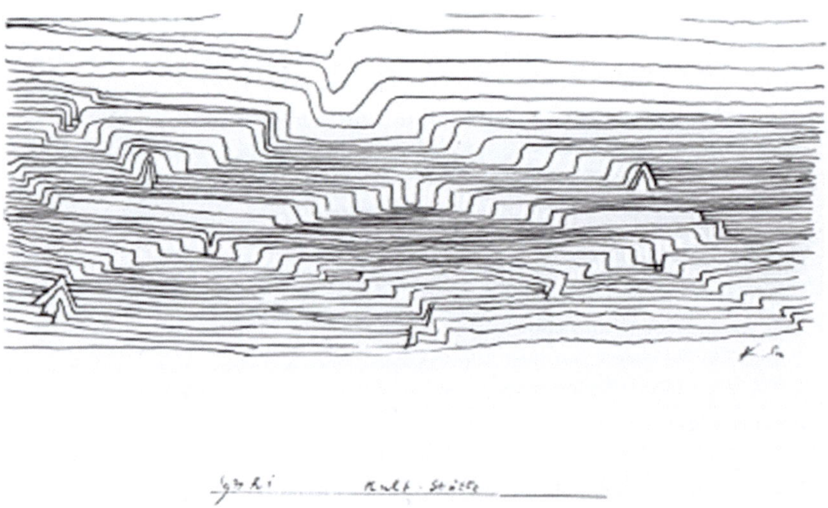

Paul Klee. *Kult-Stätte—Place of worship*. 1934, 141. Pen on paper on cardboard. 11.6 × 27 cm. Swiss private collection, deposited in Zentrum Paul Klee, Bern. *Photo* © Zentrum Paul Klee

Spaces in Space

Place of worship (1934) reruns this oscillation between spatialities to rethink the two dimensions of the surface. A subtle dialectics appears between the unilinearity of the strokes and the three-dimensional morphology of the fractal landscape that all the lines compose. This work is just one of the achievements of the artist's long research concerning the dimension of lines or colour layers, on the one hand, and the effect of a three-dimensional representation on the other. The conflicting interweaving of different perceptual modes leads to the configuration of what René Thom defines as spaces in space. This idea of "a particular sort of three-dimensional representation on the flat surface" that—as Paul Klee explains—"force[s] the third dimension into the flat plane" constitutes, in Thom's terms, one of the "means by which one space can be carried into another in a continuous fashion" (Thom 1991, p. 39). Thus, *Moved thresholds* (1929) plays on the tension between two asynchronous dynamics that nevertheless coexist simultaneously. In continuity with the quiet undulation of the horizontal layers, more agitated waves emerge in the vision, perpendicular to the gaze of the beholder. Hence, a sensation of compression results and, at once, of a three-dimensional deployment problematizing the surface. Yet each stratum of the rapids regains and fits perfectly in the layers of the calmer waves. The contrast does not manifest itself in the detail but in the perceptual grouping of the visual elements. The two configurations finally give rise to a blend of gestalts, which preserves its conflicting dimension.

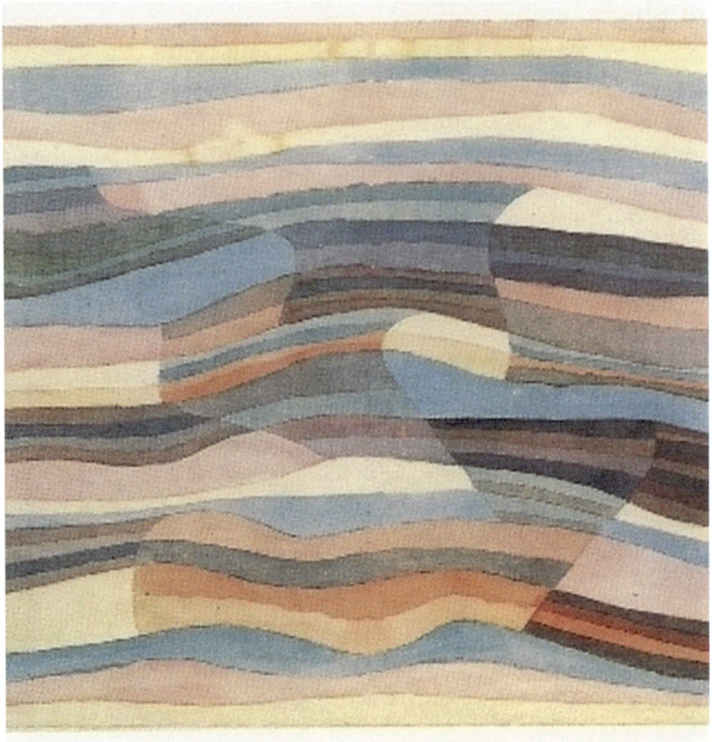

Paul Klee. *Bewegte Schwellen—Moved thresholds*. 1929, 236. Watercolour and pencil on paper on cardboard. 34 × 37 cm

In an interview, the mathematician Thom enlightened how the conception of spaces in space leads to review the concept of spatiality that tradition, since Euclidean geometry, considers as homogeneous:

> Generally speaking, the spaces I've looked at are what are called homogeneous or locally homogeneous spaces. They are also called manifolds. Ordinary Euclidean space is a manifold. The application of constraints is what produces **singularities**. For example, ruffling the sleeve of my jacket causes several folds to appear. These do not arise from some internal mechanics in the fabric. The relevant theorem is actually quite abstract: When a space is subject to a constraint, by projecting it onto another space of smaller dimension, the target space will continuously assimilate the constraint everywhere save at a certain number of points where, in a manner of speaking, it concentrates its individuality! It manifests its resistance through the presence of these singularities. The concept of a singularity allows one to embody an entire global structure. (Ibid., p. 40)

An incredible intuition leads Klee to force three-dimensional spatiality, to show how the illusionist representation is subject to the constraints of the surface. Eliminating definitively the foreshortening used in traditional figuration, he highlights all the problematic aspect of such pictorial transposition. He works tirelessly on the points of resistance, on the edges and critical elements that make the perception leap from one mode to another. *Monument on the border of the fertile country* (1929) precisely plays on the concept of thresholds that opposite configurations define. The conflict between a dominant flatness, set as a norm, and the sketch of a plastic depth, deviating from the former, mainly concentrates on a few oblique lines, summing up the maximal perceptual tension.

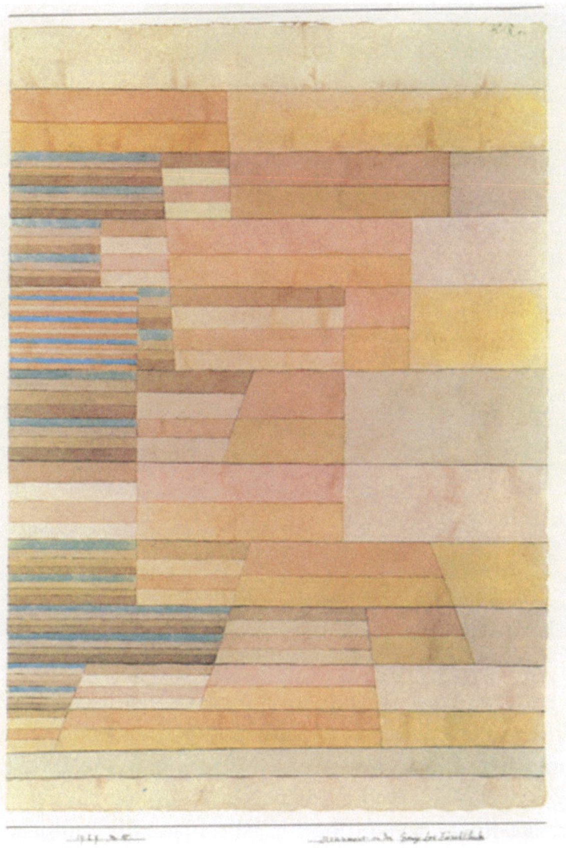

Paul Klee. *Monument an der Grenze des Fruchtlandes—Monument on the border of the fertile country*. 1929, 40. Watercolour and pen on paper on cardboard. 45.8 × 30.7 cm. Museum Sammlung Rosengart, Lucerne. *Photo* © Zentrum Paul Klee

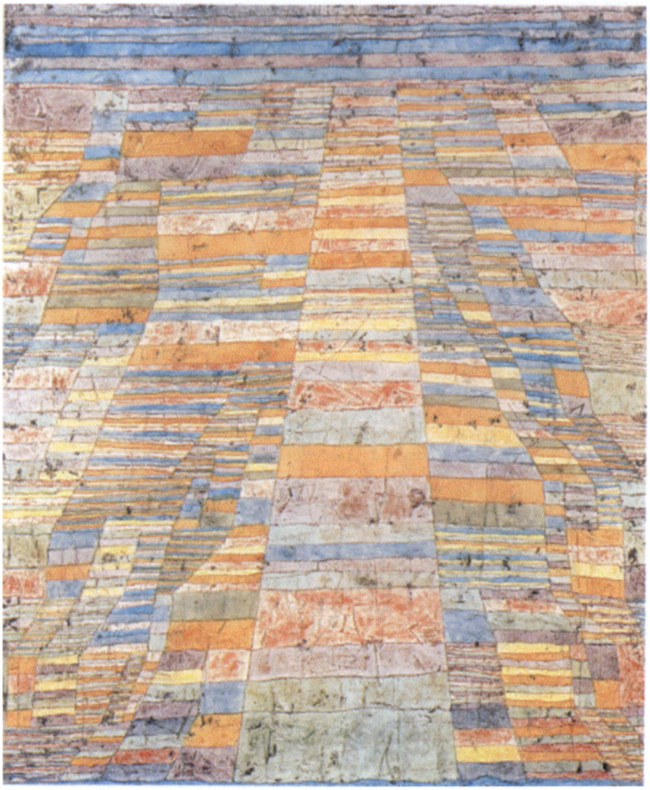

Paul Klee. *Hauptweg und Nebenwege—Highway and byways*. 1929, 90. Oil on canvas. 83.7 × 67.5 cm. Museum Ludwig, Cologne

This tension between flatness and depth appears also in another painting of the same period, *Highway and byways* (1929). The duality between horizontal layers and depth becomes here more complex. The depth is fragmented in a multitude of different perspectives, heights and gradients, bringing the pathway network back to an effect of intricate lines. In addition, the chromatically uniform layers, crossing the surface at the top and the bottom, contrast with the perspective articulation that develops between them. They encourage another perception of space: not any more the vision of the main path as climbing, but as a horizontal bridge connecting planes of the same height, at the side of which the faults of an uneven ground open out. All these perceptual possibilities fit together, overlap, and conflict with each other. The vision only stabilises weakly and is promptly put into question. Finally, even though the perspectival depth, despite its reversals, seems preponderant, an ultimate factor refocuses the attention on the two-dimensionality of the picture. The materiality of the background indeed gives homogeneity to the whole and blurs any representation, whether abstract or figurative. The significance of this materiality has properly been noticed by Boulez in a comment on another Klee's painting, *Refuge* (1930), but it also concerns *Highway and byways*:

> The preparation of the background was for him such a primordial stage of creation that he did not measure his time to obtain what he was looking for; sometimes he spent a month modifying it through various techniques. So that the result is an extraordinarily developed but at once amorphous background, that is, it has no directionality. We do not know how to look at it; or there are a thousand ways to look at it, as we would do with a cloud, but a cloud that would have been suddenly fixed, where the mobility is in the eye and no longer in the object itself. When we look at a cloud, the gaze espouses and follows the mobility of the cloud. But in front of a background by Klee the mobility depends on the way we look at it. This implies that the space is oriented according to the perceiver's will. In other words, the power of the gaze is what modifies the texture of the painting. (Boulez 1989, pp. 163–166, own translation)

Whether oscillating between figuration and abstraction, or between depth and surface, forms and amorphous matter, lines and planes, *Highway and byways* unfolds the potentiality of these contrasting attractors that dynamise its perception. It could therefore be placed in that middle area Klee defined, for instance, between the domain of the line and the domain of the surface (cf. Klee 1956, English transl.: p. 120):

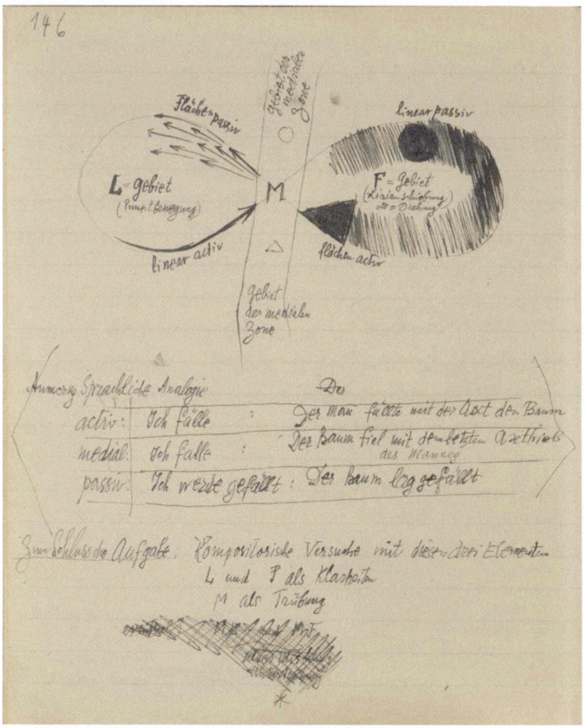

Paul Klee. Beiträge zur bildnerischen Formlehre—Contributions to the theory of pictorial form. 20.2 × 16.3 cm. Zentrum Paul Klee, Bern, Inv. BF/149. *Photo* © Zentrum Paul Klee

If generally in the pictorial forms either the line is active and the surface passive, or the surface is active and the line passive, there exist intermediate forms where neither of them seems to impose its strength. The area between the two domains concentrates all the dynamism of the structure in a point. This area where the catastrophic passage between two poles occurs becomes the energy centre of dynamic potentialities. It is certainly not by accident that in a very similar schema Klee places in this point the heart, vital centre of blood circulation (cf. ibid., p. 355). It's another way of emphasising how the dynamism between opposing poles, touching its climax in this area, make the form become a living one.

Klee himself proposes another analogous intermediate area where the entrenchment between the active and the passive turns out to be problematic. He refers to the active and passive forms of sentences in linguistics and, in this respect, detects expressions in which the verbal form looks logically undecidable.[13] The artist, having also a disposition to poetry in addition to music, repeatedly draws attention to language. He rethinks in particular the writing whose visual and graphic form lends itself to becoming a nonfigurative element for his pictures. In his criticism, Louis Marin noticed that the use of the letter in Klee's works of art gradually disappears "to give way to a writing-painting" (Marin 1970, p. 105, own translation). Actually Michel Foucault's *Ceci n'est pas une pipe* (1973) already recognised that the intercrossing of painting and writing, that is, a system of representation by resemblance and a system of reference through signs, gave birth to a space *other* with respect to these two domains. More recently, Per Åge Brandt has re-read this dialectic that the "*form* [...] keeps having with the *image* (in the «representation») and with the arbitrary *signifier* (in the «symbolisation»)" (Brandt 1991, own translation). Brandt thus discerns "a perfect instability" (ibid., own translation) that plunges the form into the abysm between two statuses of the sign: the icon and the symbol. The two-dimensional form would therefore be taken between two different attractors, the representation that refers to the three-dimensional figuration and the symbolism of the word, which is on the contrary one-dimensional. According to the author, this space is organised as per the model of "a cusp, in catastrophe theory" (ibid., own translation):

[13] Here is a translation of the examples he gives (Klee 1956, English transl.: p. 120):

Active: I fell: The man felled the tree with the axe.

Middle: I fall: The tree fell with the man's last stroke.

Passive: I am felled: The tree lay felled.

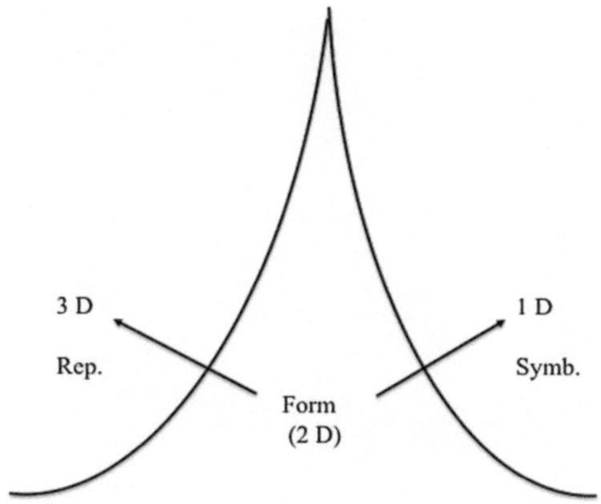

Per Åge Brandt. Graph. *Source* Brandt (1991)

Brandt's reading, inspired by Thom's catastrophe theory, has the advantage of grasping some of the semantic aspects that articulate Klee's work. The meaning would thus be given, *infinitely* as the author explains, through this mise en abyme of the form oscillating ceaselessly between two contrasting signifying values. In fact, though in some paintings there is still a distinction between the pictorial and the graphic and yet a kind of hierarchy appears (e.g. *Ad marginem*, 1930), in many others, these borders are exceeded: writing becomes pictogram (*Egypt destroyed*, 1924), painting becomes graphics (*Ara (Cool breezes in a garden in the torrid zone)*, 1924). This is not without recalling the return to the spatial structure of a childish art, as Fernande Saint-Martin observed regarding Klee (cf. Saint-Martin 1980, p. 144). However, Brandt's reading seems to simplify somehow the dynamism of Klee's art. Iconicity and symbolisation are just two of the issues the formation of form involves. It would be necessary to multiply the attractors that define this space, to add for instance the tension between abstraction and figuration, between several representational modes, between opposite perspectives and points of view, between contrasting spatial constructions, and still to take into account the relations between the other parameters that Saint-Martin's topological syntax prompts to find out: chromaticism, tonality, texture, morphology, vectoriality, dimension. In a nutshell, this space in abysm very likely turns out to be a veritable space of infinite dimensions. In *Florentine Residential District* (1926), the line becomes calligraphic; but in interaction with the colour areas, the script initiates a landscape, bits of figuration, architecture: houses, façades, spaces, vegetation. Horizontal vectoriality—from left to right, from top to bottom, as it may be tempting to say—becomes complex in several directions; it generates surfaces, planes which organise themselves in higher dimensions. The more the observation refines its visual analysis tools, the more the structure reveals to be rich in conflictual blends. Subjugating these higher dimensions to the constraint of the surface, the sense of Klee's pictures definitely fibrillates.

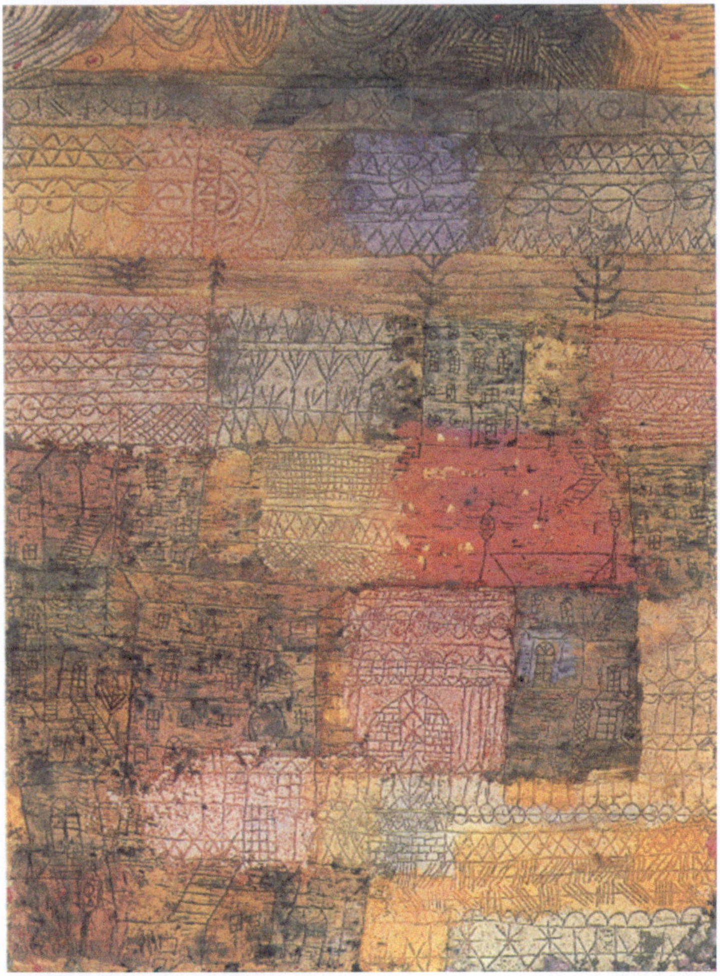

Paul Klee. *"Florentinisches" Villen Viertel—Florentine Residential District*. 1926, 223. Oil on cardboard. 49.5 × 36.5 cm. Musée national d'art moderne, Centre Pompidou, Paris

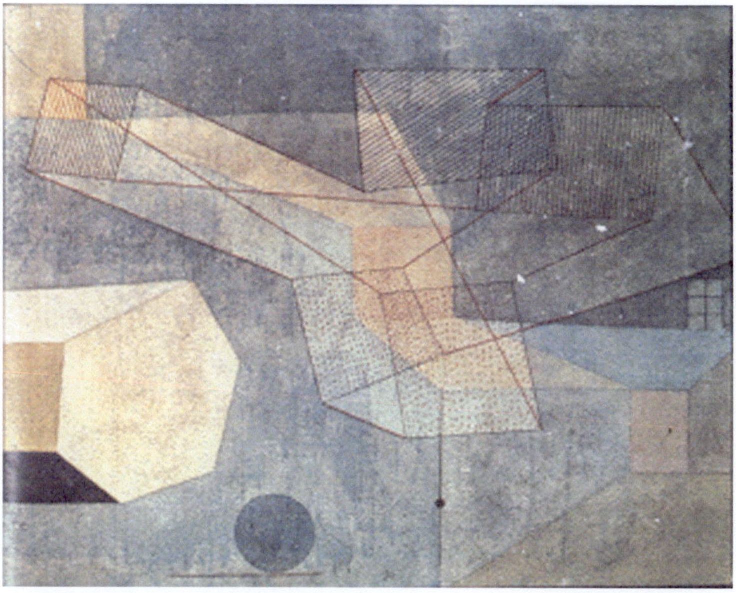

Paul Klee. *Segelflug-zeug—Glider*. 1930, 181. Oil on cotton, original frame. 34.4 × 43.5 cm. *Photo* © Zentrum Paul Klee

Representation in Catastrophe

In several works of art, the objects of representation undergo a deformation, a transformation in their dimensions. *Glider* (1930) depicts geometrical elements, which develop in an undecidable spatial arrangement. Compressed like the sleeve of his jacket of which Thom speaks about in order to illustrate folds, this complex volumetry flattens, changes to adhere to the surface of the canvas. But perceptual tensions, points of resistance and singularities reappear, displaying the entire problematic aspect of such a two-dimensional inscription. In other cases, Klee uses the interposition of a glass disc to deform objects. The same process was at the origin of the linear perspective: once the glass disc is fastened vertically, facing a fixed eye, in front of the objects to be reproduced, their essential features can be drawn (cf. Klee 1956, English transl.: p. 149). Nevertheless, with Klee the accuracy of the process turns against itself, the disc tilts, the proportions of the projected object change, the image looks distorted.[14] "Such images – makes clear the artist – are not even objectively or logically wrong", they are "physiologically wrong" (Klee 1956, English transl.: p. 149). What interests him indeed is the conflict between the accuracy of the process and the perceptual effect, that third state between "*physis*" and "*psyche*" wherein he situated himself (Klee 1957, English transl.: p. 125). *Scholar* (1933) looks like having been realised by means of a tilted disc: the figure lengthens, the elongation transforms perception, the personage acquires an unusual air reinforcing the expressiveness of his face.

[14]Cf. also the description in Boulez (1989, pp. 64–65).

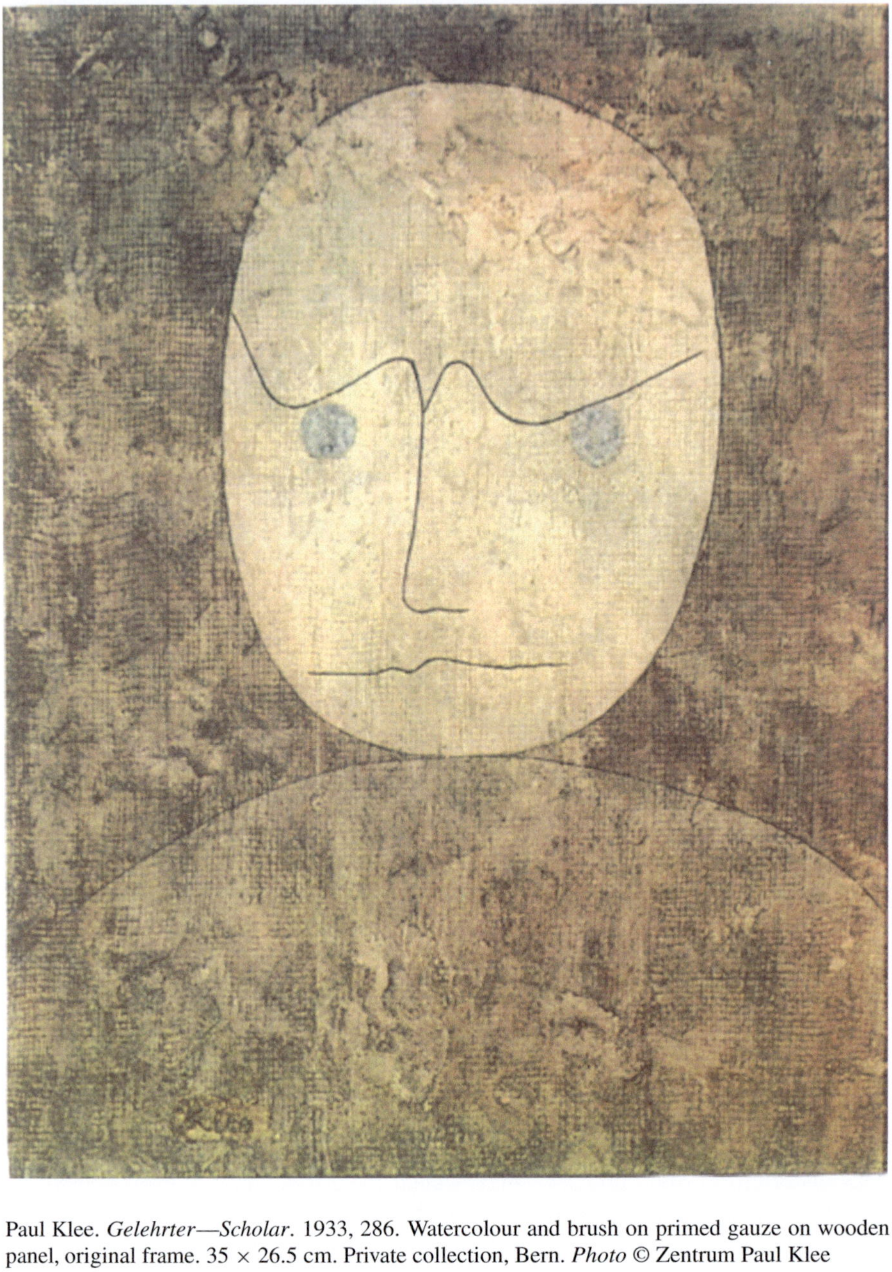

Paul Klee. *Gelehrter—Scholar*. 1933, 286. Watercolour and brush on primed gauze on wooden panel, original frame. 35 × 26.5 cm. Private collection, Bern. *Photo* © Zentrum Paul Klee

Another example, brilliantly illustrated by Boulez, shows how the painter conceives the relation between two different domains. Commenting on one of his students' exercise, Klee reflects upon the possible interaction between a line and a circle in such a way that not a simple composition or a fusion will follow, but a transformation of an element in relation with the other:

> he then examines: the possible relations between those two very simple but heterogeneous elements that are the circle and the straight line; the figures of the struggle that the fluid and the solid natures can engage; their implicit forces; how to imagine their encounter, antagonism and conjugation.
>
> […] And as a consequence of such a study, which may seem very dry and theoretical, Klee transfigures this student exercise into a remarkable watercolour of intense poetry: *Physiognomischer Blitz*, *Physiognomic Lightning*, which I would willingly translate with *Lightning crossing a face*. The problem is forgotten, the invention has taken advantage of it […]. The result is both unpredictable and profoundly logical. (Boulez 1989, pp. 133–134, own translation)

In *Physiognomic Lightning* (1927), the straight line does not merely cross the circle, is destructured, fragmented by it. A space of transformation takes place, at the same time modifying the circle: while crossing the latter, the straight line marks the critical points of discontinuity of the face. The transformation is catastrophic, straight line and circle are no longer discernible, the capture of the one by the other changes the physiognomy of the latter. The representation attains the point of maximum instability, hence the effect of immediacy perceived in the picture.

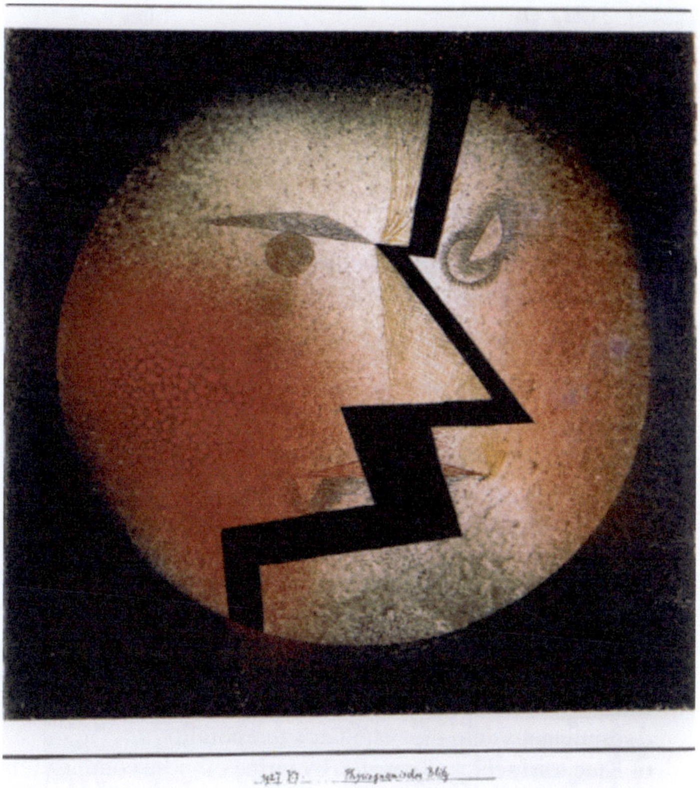

Paul Klee. *Physiognomischer Blitz—Physiognomic Lightning*. 1927, 247. Watercolour on paper on cardboard. 25.5 × 25.5. *Photo* © Zentrum Paul Klee

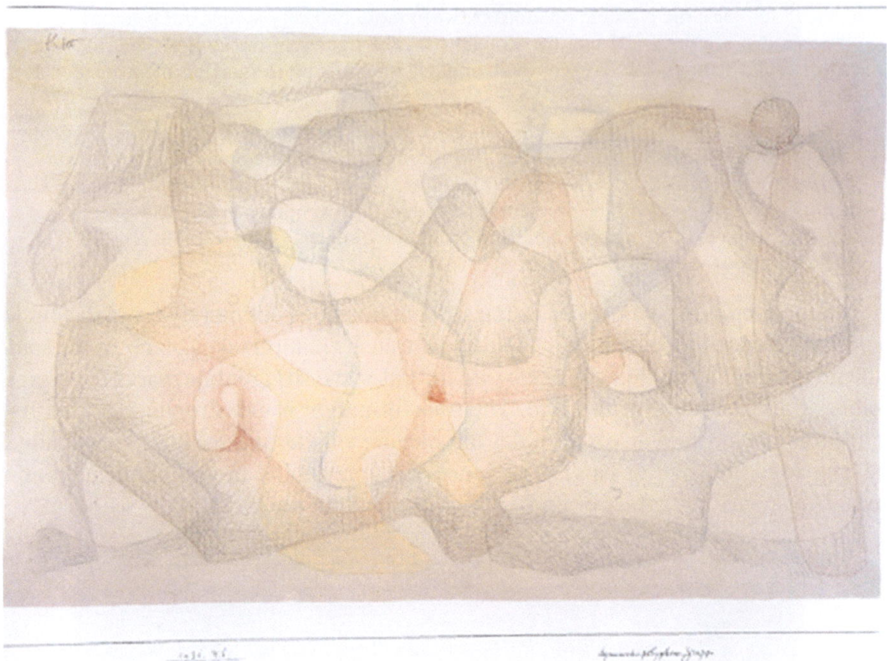

Paul Klee. *Dynamisch-pholyphone Gruppe—Dynamic-polyphonic group*. 1931, 66. Pencil and coloured pencil on paper on cardboard. 31.9/31.1 × 48 cm. Swiss private collection, deposited in Zentrum Paul Klee, Bern. *Photo* © Zentrum Paul Klee

Dynamic-polyphonic group (1931) displays an even more catastrophic representation, obviously referring the word *catastrophic* not to a judgment on the aesthetic value of the work, but to a mode of representation of the space in painting. The emerging perceptual effect conveys the impression of a complex blend of several bodies. However, not only does no representation appear, but hardly any form can be delimited. What in the detail creates tensions between lines and, consequently, between differentiating planes, in the whole is lost in an indistinguishable amalgam. These lines do not close a figure on a background, or on other figures, but remain open; they mark sometimes the empty interior, sometimes the outer contours, of a space apparently out of dimension. The eye slides from one part to the other, following the course of the lines or the colour extents, but it inevitably falls into the trap of a plane that turns on itself or metamorphs into another plane. These differentiated paths, these loops, superpositions, fusions thus redefine, through subtle variations of nonfigurative elements, the topological space of the representation. The gaze follows the dynamics of a formation that never ends in a finished form and which extends the polyphonic simultaneity into the beholder's temporality.

Almost all of Klee's art seems to develop by virtue of these complex relationships arising between visual elements. The dynamism of the form and, in general, of the modes of expression gives a first meaning to the picture. This meaning, which is no less cultural, imbued with a certain metasavoir, is rooted in and builds on perception itself. It is in visual perception that the unstable equilibrium, morphodynamism and, indeed, all the tensions encountered between incommensurable, yet topologically coexisting dimensions are defined. The empathic projection, the process of embodiment which animates objects in painting, whether they are figurative or not, further confirms this idea. A first core of meaning emerges visually, through cognitive processes allowed by this medium. Of course, the quest for meaning may also be pursued in other directions, and polysemy will not fail to emphasise the artistic value of Paul Klee's painting. But it is noteworthy that, before joining historic-artistic connotations or even psychoanalytic ones as Saint-Martin's topological approach proposes, a semantics can be deployed thanks to a visual reflection. For this purpose, metadiscursive tools have been borrowed from the theory developed in the wake of René Thom,[15] because its diagrams suggest a possible application to the questions under consideration, even if the complexity of both his theory and the numerous parameters put on the line by the painter refrains from its hasty use.

Finally, the hypothesis that a topological syntax can be combined with a dynamic semantics leads to reconsider the relationship that unites the plane of expression with this type of content. A hypoiconic rapport, a kind of inner motivation determining the relation of visual signs to each other, seems to persist also when the sign becomes figurative. To that extent, it would be possible to reframe the question of the semi-symbolic put forward by Jean-Marie Floch and Algirdas-Julien Greimas, then pursued by Thürlemann (1982) in his analysis on Klee. From a viewpoint wherein the meaning is not any more structurally given as the generative semiotic perspective intended, but constructed by interpretations, the concept of semi-symbolism risks becoming incongruous. The concept in fact claimed that a deep structure regulated the semiotic function, that is to say, not an arbitrariness of the sign or a term-to-term correspondence between the syntactic and semantic planes, but conformity between the categories of the plane of expression and the categories of the plane of content. Rethinking the semi-symbolism within the process of emergence of meaning through semiosis is then equivalent to detecting a rapport of inner motivation organising the meaning. This rapport would rather arise at a primary level, which can be defined as hypoiconic, since it concerns in any respect the perceptual moment of the sign. Even if this rapport also holds in subsequent stages, when for example the sign conveys the idea of resemblance, it can simply originate in the perception of the picture. This is clearly an outcome of the present analysis, in which the topological organisation of Klee's works of art appears strictly related to their semantic dynamics. The relations between the various nonfigurative and figurative signs seem indeed to be motivated by the catastrophic encounters they generate.

To conclude, a new value can be assigned to the term *catastrophe* that titles some works of the artist, works which his son Felix Klee has maybe inappropriately assimilated to the series devoted to the war (cf. Felix Klee 1960, French transl.: p. 95). *Catastrophe J four* (1925) pertains to Paul Klee's conflicting conception of the pictorial space. Anthropomorphic bodies slide into the emptiness of the surface, which, by its importance, seems to prevail and transform them as regards their figuration. This image hovering in the work of art appears to materialise, as a symbol, the "flying man" to whom Klee thought in 1902 when, having not yet found his style, in

[15]For his seminal theory, please refer also to Thom (1977) and Thom (1980).

his diaries he considered the possibility of forcing the representation to enter the spatiality of the painting.[16]

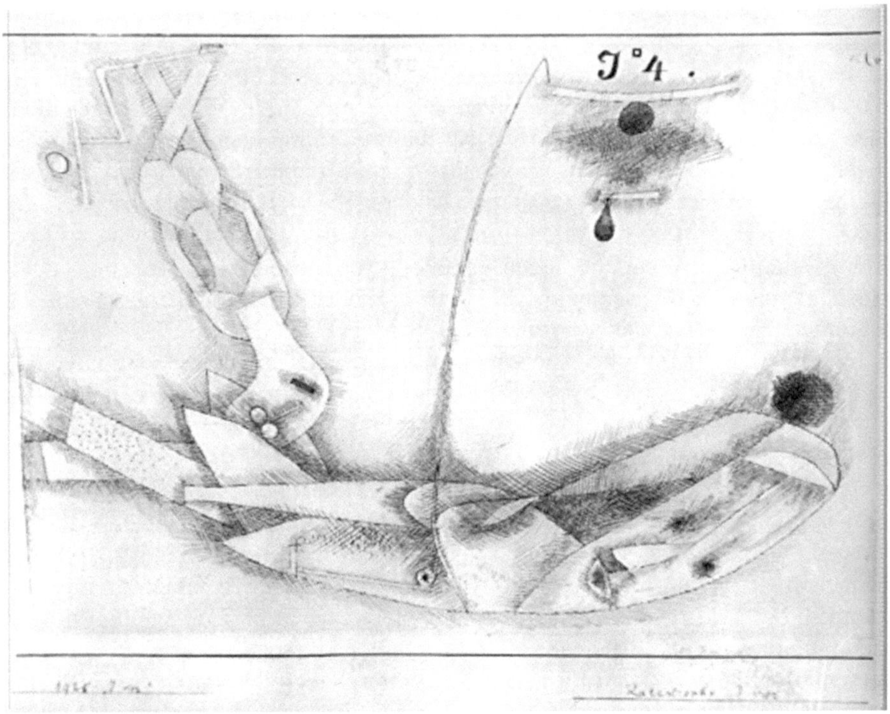

Paul Klee. *Katastrophe J vier—Catastrophe J four*. 1925, 184. Pen on paper on cardboard. 15.9 × 23.5 cm. Private collection, USA. *Photo* © Zentrum Paul Klee

References

Boulez, Pierre: Le pays fertile Paul Klee. Gallimard, Paris (1989)

Brandt, Per Åge: La beauté de l'abîme. Paper read at the Kunstakademiet on the 11 May 1991 and transmitted by the author (1991)

Brandt, Per Åge: La Charpente modale du sens. Pour une sémio-linguistique morphogénétique et dynamique. Århus University Press and John Benjamins Publishing Company, Århus, Amsterdam, Philadephia (1992)

Caliandro, Stefania: Empathie, signification et art abstrait. Visio, 4, 2, pp. 47–58, summer (1999)

Caliandro, Stefania: Le point gris de Paul Klee. Visio, 6, 4, pp. 21–41, winter (2001–2002)

Foucault, Michel: Klee, Kandinsky, Magritte. In: Foucault, M. Ceci n'est pas une pipe, pp. 39–58. Fata Morgana, Montpellier (1973)

[16]Cf., *supra*, the quotation from Klee (1957, English transl.: p. 124).

Klee, Felix: Paul Klee, Leben und Werk in Dokumenten, ausgewählt aus den nachgelassenen Aufzeichnungen und den unveröffentlichten Briefen. Diogenes, Zurich (1960) [French transl.: Paul Klee par lui-même et par son fils Félix Klee. Les Libraires Associés, Paris (1963)]

Klee, Paul: Das bildnerische Denken, collected and edited by Jürg Spiller. Schwabe & Co., Verlag, Basel, Stuttgart (1956) [English transl.: Notebooks. Volume 1. The Thinking Eye. Lund Humphries, London (1961)]

Klee, Paul: Tagebücher von Paul Klee 1898–1918, edited by Felix Klee. Verlag M. DuMont Schauberg, Cologne (1957) [English transl. authorized by Felix Klee: The Diaries of Paul Klee, 1898–1918. University of California Press, Berkeley, Los Angeles, London (1964)]

Klee, Paul: Unendliche Naturgeschichte, collected and edited by Jürg Spiller. Basel, Stuttgart: Schwabe & Co., Verlag (1970) [English transl. Notebooks. Volume 2. The Nature of Nature. Lund Humphries, London (1973)]

Marin, Louis: Klee ou le retour de l'origine. Revue d'Esthétique, January (1970). New edition with the title: Klee ou le retour à l'origine. Études sémiologiques. Écritures, Peintures, pp. 101–108. Klincksieck, Paris (1971)

Thom, René: Modèles mathématiques de la morphogenèse. Recueil de textes sur la théorie des catastrophes et ses applications. Union générale d'éditions, Paris (1974). New edition, revised and augmented: Christian Bourgois Éditeur, Paris (1980) [English transl.: Mathematical Model of Morphogenesis. Ellis Horwood Limited, New York, Brisbane, Chichester, Ontario (1983)]

Thom, René: Stabilité structurelle et morphogenèse. Essai d'une théorie générale des modèles. New edition, revised and augmented: InterÉditions, Paris (1977) (first edition: 1972)

Thom, René: Parabole e catastrofi. Il Saggiatore, Milan (1980) [French transl. revised and completed by the author: Paraboles et catastrophes. Entretiens sur les mathématiques, la science et la philosophie réalisés par Giulio Giorello et Simona Morini. Flammarion, Paris (1983)]

Thom, René: Prédire n'est pas expliquer. Second edition revised and corrected. Éditions Eshel, Paris (1991). New edition: Flammarion, Paris (1993) [English transl.: To Predict is Not to Explain, edited by Tsatsanis, S.P. http://www.fermentmagazine.org/Stories/Thom/Predire.pdf]

Thürlemann, Felix: Paul Klee. Analyse sémiotique de trois peintures. L'Âge d'Homme, Lausanne (1982) [one of the essays, «Blumen-Mythos»—1918, has been translated in Italian: Paul Klee: analisi sémiotica di Blumen-Mythos—1918. In: Corrain, L., Valenti, M. (eds.) Leggere l'opera d'arte. Dal figurativo all'astratto, pp. 107–131, ill. 13–14. Esculapio, Bologna (1991)]

Saint-Martin, Fernande: Les fondements topologiques de la peinture. Essai sur les modes de représentation de l'espace, à l'origine de l'art enfantin et de l'art abstrait. Hurtubise HMH, Montreal. Consulted edition: 1989 (1980)

Wildgen, Wolfgang: Process, Image, and Meaning: A Realistic Model of the Meaning of Sentences and Narrative Texts. Benjamins, Amsterdam (1994)

Wildgen, Wolfgang: De la grammaire au discours. Une approche morphodynamique. Lang, Bern (1999)

Wildgen, Wolfgang: Kurt Lewin and the rise of cognitive sciences in Germany: Cassirer, Bühler, Reichenbach. In: Albertazzi, L. (ed.) The Dawn of Cognitive Science. Early European Contributors, pp. 299–332. Springer, Dordrecht (2001)

Worringer, Wilhelm: Abstraktion und Einfühlung (1908) [English transl.: Abstraction and Empathy. A Contribution to the Psychology of Style. International Universities Press, New York (1953)]

Chapter 3
Vibration as a Morphogenetic Force of Creation. Architectonic Dynamics in the Work of Peter Eisenman

> Many of you will know the pretty experiment with sound figures. First spread a layer of fine sand on a thin plate of wood or metal. Then draw a violin bow across an edge, making the plate vibrate. This vibratory impulse is the heart of the matter. It causes the sand to arrange itself in a corresponding rhythmical order. First, in other words, the vibratory impulse, the will or need for living action, then the transformation into a material event, and lastly its visible expression in the form of newly rearranged material.
>
> (Klee 1970, English transl.: p. 44)

With these few words, simple and rich at once, Paul Klee explained to his pupils at the Bauhaus his *plastic thinking* and how, in art as in nature, matter can be seized by an expression of life. Form is interesting insofar as it manifests the law of its becoming, the dynamism of its formation, the morphogenesis and the entelechy that determine it: the vibration imparts an impulse, a rhythm, a structuration to the matter, introducing an organic dimension into the form, beyond any mechanical causality. The teleonomic arrangement of the form is then apprehended as "visible expression" of "the will or need for living action". I will not focus here on the artist's aesthetic but, through his writings on art theory, I will try to apprehend the question of vibration from a semiotic viewpoint. As paradoxical as it may seem, this phenomenon concerns not only the so-called arts of time, such as music, poetry, theatre, cinema and dance, but also some visual works of art, including architectures, in which the movement is, so to speak, condensed in a given form or a still picture.[1]

This question meets the diagrammatic conception animating Peter Eisenman's architecture. Despite the *metaphysics of presence* that, according to his reading of Jacques Derrida, is characteristic of this art (cf. Eisenman 1990, new edition: p. 4),

[1] A French version of this paper has been published in the Canadian academic journal *Cygne noir* (Caliandro 2013).

S. Caliandro, *Morphodynamics in Aesthetics*, Lecture Notes in Morphogenesis,
https://doi.org/10.1007/978-3-030-29631-5_3

his realisations and designs of architecture show—and go beyond—the inner process (the semiotic and architectural functions) and the pre-existing elements (the actual site, the historical and cultural heritage, etc.) that engendered them. Eisenman's theoretical and architectonic approach is particularly developed in this respect, since it includes an unexpected definition of forms in becoming, for the *superposition* (term that will need to be specified) of variations and variants adopted in the creation (cf. Eisenman 1998). The concept of vibration then hybridises and enhances itself with the idea of *interference* (cf. Eisenman 2000, new edition: p. 96), implying in a way a dynamisation of both the object of the gaze and the gaze that the beholder directs to the object. In other words, vibration concerns not only the work of art but also, and perhaps even more, its moments of creation and reception.

By considering what kind of sign the vibration could refer to according to Charles Sanders Peirce's theory, I will expound the difficulty of simply including this phenomenon in the category of indexes,[2] traces or imprints. In addition, I will discuss how, in a deconstructionist perspective, vibration has became a means to attain the de-structuring of the architectural space. However, I will not dwell on the last point, which is more obvious because of the close ties, also historic, between architectural and philosophic postmodernisms, and rather put forward a morphodynamic intelligibility of the work of art, wherein the vibration impressed in the creation process continues to be transmitted in its psychophysic perception.

A Morphodynamic Conception

As Paul Klee wrote, the "vibratory impulse is the heart of the matter"; by it the matter, modulated through oscillation, takes form according to a precise structure. This structure is not conceived as rigid or inanimate and, therefore, the artist prefers to symbolise it by a wavy line instead of a mensurable straight one (Klee 1970, English transl.: p. 45). Depending on its disposition and propensity for motion, the matter then enters the stream and adapts itself to "a living idea" (ibid., p. 49). The force—this "creative power" which "cannot be perceived with the same senses as the more familiar kinds of matter" (ibid., p. 63)—can only be visualised through formation and the process of continuous evolution it generates. It is at once "ultimately mysterious", since we are ourselves charged with it "down to our subtlest parts", and "a form of matter" that can be apprehended by the way it models the sensorial phenomenon. It is of interest to note that, for the artist, this force must be thought as immanent—"a form of matter"—"permeated with the matter"; while affecting the latter, it "affects us deeply" (ibid.). It gives the measure of the apperception of a sensorial intelligibility, since this force passes through and manifests in matter without transcending it.

[2]In the French translation, the term *indice* (clue) has often been employed, to refer to what other authors, in the wake of the North American tradition (Umberto Eco, Rosalind Krauss, etc.), prefer to name *index*. In the latter case, the word *indice* (e.g. the trace of a crime) indicates rather a specific type of *index*.

All of Klee's plastic thinking is based on the articulation of forces, antagonistic and dynamic tensions that are not reducible to an abstract system but pertain to an intrinsic organisation of forms. In this sense, it can be assimilated to the same morphogenetic conception that led Claude Lévi-Strauss to distinguish the idea of structure from the concept of system. In a very enlightening synthesis, Jean Petitot has demonstrated the genealogy, the essence and the horizon of the structuralism, by placing this in the epistemic line that, from Goethe to D'Arcy Thompson and René Thom, has worked at a naturalisation of phenomenology and the perceptual structures. From this perspective, Petitot emphasises how the anthropologist wanted to distance himself from a certain structuralist formalism, to find affinities with theories of nature and natural sciences focusing on morphologic dynamics.[3] These sciences—"theory of catastrophes and bifurcations of attractors in nonlinear dynamic systems, theory of critical phenomena and symmetry breaks, self-organisation theory, self-organised critical states, nonlinear thermodynamics and dissipative structures, etc."—have now formulated physic-mathematical models for a qualitative understanding of the complexity of non-elementary systems (Petitot 1999, p. 115, own translation). In a more intuitive way, although informed by non-Euclidean geometries, Klee seems to have brought together physic-acoustic considerations from material science, studied since Ernst Chladni, and some contemporary research on Gestalt psychology to think jointly, as Petitot nowadays recommends to do, "not only geometry and forces but geometry, forces and forms" (ibid., p. 101).[4]

Vibration is both a force and forms, the latter being governed by an organising geometry strictly dependent on the type of propagation in a given material. The geometry of forms manifests in their topological arrangement. Through the perception of forms, their oscillation and their spatial distribution, a force can indeed be apprehended. To be well seized, vibration signs need to be integrated into a semiotic function, as Umberto Eco wrote concerning signs such as fingerprints, symptoms and cues (cf. Eco 1975; Eco 1972, French transl.: pp. 165–183 with table p. 174; and Eco 1992), because in most cases their genesis is not immediately perceivable, but reconstituted by inference or interpretation.

Peter Eisenman: An Architecture in Becoming

In his writings, Peter Eisenman constantly reflects upon an architecture that keeps tracks of its becoming and that displays, even in the final realisation, its evolution process. In reaction to modernist functionalism, the architect supports the idea that

---

[3]Jean Petitot quotes in particular Claude Lévi-Strauss from *De près et de loin*: "I will even say that all the errors, all the abuses committed on or with the notion of structure come from the fact that their authors did not understand that it is impossible to conceive it separately from the notion of transformation. The structure cannot be reduced to the system: a set composed of elements and relations that unite them." (Petitot 1999, p. 98).

[4]This author has recently resumed the perspective linking Claude Lévi-Strauss's structuralism to Goethe, D'Arcy Thompson and René Thom (cf. Petitot 2010–2011).

simple and stable forms are only the reduction of more or less complex systems—as evinced by earthquakes, landslides and tsunamis—of which they barely constitute a suspension (cf. Eisenman 2004).

Reader of Noam Chomsky, then of Michel Foucault, of Deleuze and Guattari and, above all, of Jacques Derrida, he utilises Peirce's but also Deleuze's notion of diagram to elaborate, for example, in the seventies, variants more and more articulated and reworked of the architectural basic form of a house, the cube. Referring to Wittkower's schemas concerning Palladian villas, as well as to Le Corbusier's Modulor, Eisenman distinguishes two different uses, one analytical or explanatory and the other as a form generator. The first use is called "postrepresentational": the diagram is outlined after the realisation and helps to understand the relationships, including spatial, inside the architectural work (Eisenman 1998, new edition: p. 88). Since it does not give a point-by-point representation of the object, the diagram pertains to the hypoiconic dimension recognised by Charles Sanders Peirce, for whom the diagram indicates, in its own parts, the internal relations between the parts of its object.[5] But there is a second way of considering the diagram: as a sign of a thinking still in motion, which includes traces of becoming, that is to say the possible variants that the same architecture could have had. This is a more projective dimension that refers to Peirce's conception of diagram as a tool for theoretical reasoning and thoughts, but which has undoubtedly been influenced by Gilles Deleuze's interpretation of this notion. Thus, the diagram, as a hypoicon, "«unlocks areas of sensation»", it works the moment prior to any figurative stabilisation (the *figural* according to Jean-François Lyotard) "to introduce «possibilities of fact»" (Deleuze 2002, English transl.: pp. 101–102).[6] The diagram carries "an architecture of traces", becomes medium for creation, insofar as it refers to the history of the project and let the project's dynamics reappear (Eisenman 1998, new edition: p. 89). Its designing dimension, while not excluding the analytic moment of the diagram, builds on the combinatory, almost tautological splitting of his first creations to enhance variability as an architectonic force and matter.

Inspired by the Derridian text on Freud's "Note upon de «Mystic Writing Pad»", Eisenman develops the idea of a continual rewriting, where previous writings cannot truly be erased, because, like in the magic pad which exemplifies for Freud the functioning of the unconscious, they remain engraved in the wax table under the more superficial writing sheet. Similarly, he thinks of a stratification of architectural writings, which the diagram let, however, resurface and which, through it, are transmitted to the project, then to the building. It is worth noting that, as the architect states in his aesthetic declarations, these traces are not and cannot simply be integrated into a solution aiming at dialectically resolving figurative discordances. In this sense,

[5] In distinguishing images, diagrams and metaphors in the typology of hypoicons, Charles Sanders Peirce calls diagrams "those which represent the relations, mainly dyadic, or so regarded, of the parts of one thing by analogous relations in their own parts" (Peirce 1931–1958, 2.277).

[6] Gilles Deleuze uses the notion of diagram from a statement of Francis Bacon (cf. Deleuze 2002, English transl.: note 3 p. 184). In the abovementioned quotations, he cites, respectively, Francis Bacon and Wittgenstein. This notion is elaborated several times in his writings (cf. De Landa 1998). Cf. also: Lyotard (1985).

he distances himself from Gilles Deleuze's notion of superimposition, which Eisenman understands as a vertical layering, while for him the "*superposition*" implies an overprinting of layers, hence intermingled and indissociable, or, in his own words, "a coextensive, horizontal layering in which there is no stable ground or origin, where ground and figure fluctuate between one another" (ibid., p. 90). This interaction between figure and ground or between several figures takes to the extreme the question of transparency, whose ambivalent and equivocal feature Colin Rowe and Robert Slutzky have put forward, since it entails a simultaneous perception of different spatial localisations (cf. Rowe and Slutzky 1963).[7] For Eisenman, this leads to a highly articulated architecture, wherein the internal process of the project meets other elements prior to its design and pertaining to the site.

His *Guardiola House* in Santa Maria del mar (Cadiz, project not realised, 1988) is emblematic of the morphodynamism he wants to inscribe in the architecture, in resonance with the forces and fluctuations proper to the place. As shown by the drawings in plan, elevation and perspective and the overall model, the L-shaped volumes, which are typical of many Eisenman's architectures, reproduce in this house the oscillation movement transmitted by the sea to the beach and to the landscape. Just as the sand is moulded by the waves and acquires an undulating morphology similar to that of the sea, so the architecture, for Eisenman, must show the traces of the vibration traversing the environment. In other works, geotectonic and orographic dynamics (*Cidade da Cultura de Galicia* in Santiago de Compostela, in progress since 1990) or, also, dynamics related to the urban and politic history of the place (*Bibliothèque de L'HUEI*, in Geneva, 1997; *Memorial to the Murdered Jew of Europe*, Berlin, 2004) are conducive to generating forms that interact with the initial diagram and participate in his rewriting. It is important to point out that these forms are only partially assimilable to footprints or traces of pre-existing elements, since they are not simply shaped on the iconic appearance of these elements and are not, strictly speaking, clues of these experiences. They rather reactivate the dynamics produced by these experiences, wherefrom the forms hence results. Thus, even if Eisenman does not express in terms of vibration, the design of his *Guardiola House*, both in its appearance and structural implantation, appropriates—by reproducing them—the fluctuations and oscillations of the site, although no physical contiguity or direct action of these forces has acted on the building. Against a merely formal valorisation of *morphing* techniques, which the architect may now use, Eisenman encourages to appreciate the smallest details, like interstitial spaces, which characterise the forms generated by the reactivation of these dynamics and which give evidence to the internalisation of these forces into the work of creation (cf. Eisenman 1999, new edition: p. 202).

In this respect, another notion is essential to understand his approach: simultaneously to the development of a diagrammatic thinking in architecture to which he contributes, Peter Eisenman does not hesitate to submit his diagrams to "a blur-

[7] In regard to this adjacency, Eisenman was in contact with Colin Rowe. The question of transparency in painting and architecture, together with and partially echoing the concept of opacity, studied on several occasions by Louis Marin, deserves a thorough analysis that cannot be done here.

ring", to a moving action, a confusing shift that makes the "figural" re-emerge in the form/function of the first project diagram (Eisenman 1997, new edition: pp. 59 and 63). Whether defined as the jitter between the figure and the ground, as the in-between or as the interstitial process between two possible forms, the use of this movement generating blurred images is a way to reintroduce in architecture "forces and flows which could be multiple, reversible, and deformative rather than linear and transformative" (ibid., p. 57).[8] Blurring is, therefore, a process that contributes to fuzziness, to the duplication or displacement of forms, by reviving the impression of a work traversed by a vibration or a basic tremor. It produces "a blurred condition" where the relations between form and content and between site and programme are questioned through "interference" (Eisenman 2003, new edition: p. 112). By this way, the process of blurring opens "new possibilities for the figural", it adds "another layer" to the strata of the diagram, by making the different moments of the internal process of the diagram interact with the other dynamics previous or prior to the development of the project (Eisenman 1998, new edition: p. 94). Thus, especially in his *Guardiola House* and projects worked since the 1990s, the dynamics outcoming from the site dislocate, deform and duplicate, without ever repeating them as they are, the elements that the architectural thinking establishes.

This is a conception that moves, first of all, the point of view, a vibration that destabilises the subject and would like to transmit to the latter the perception of a space "as a matrix of forces and sense" (Eisenman 1997, new edition: p. 69). It aims to create an "affective" space that involves the body as well as the eye and the mind (ibid.). In this respect, Eisenman distinguishes between the "affect" and the "effect" that the architecture should produce: his work does not, or not only, want to meet utilitarian requirements and realise a function, that is to produce effects in society according to a modernist logic, but solicit an affect within a physical environment (Eisenman 1992, new edition: p. 20).

From Forms to the Creative Body

The traces which Eisenman discusses and on which his works develop bear a memory that not only arises from the project but also from its mode and conditions of realisation (cf. Eisenman [1994] 2007, p. 82); they contribute to defining the "plastic mold" of his architectures (Eisenman, 2004, new edition: p. 147). Yet, one may ask how these forms, produced through diagram, blurring and their interference, give rise to the perception of a vibration traversing the architecture.

The question does not fundamentally differ from that put forward by Umberto Eco in his *Trattato di semiotica generale*, analysing the operations carried out to

[8]Taking into consideration his use of the concept of *figural*, this movement may be defined, in my opinion, as of a "transformer", in the sense given by Jean-François Lyotard in his essay on Marcel Duchamp (Lyotard 1977, new edition: 2010), for the conversion continuously operated, in Eisenman's work, between the potential and the architectural realisation.

produce and interpret signs. As the semiotician stressed, his synthetic table does not aim at a classification of "types of signs" as in traditional typologies, but of "modes of producing sign-functions" (Eco 1979, p. 217; Eco 1972, French transl.: p. 172 and table p. 174). Imprints, traces and clues are defined insofar as they induce an activity leading to the recognition of their origin. Assembled, as such, within indexes and integrated in a semiotic function that defines them as signs, imprints distinguish nevertheless from symptoms and clues for the type of referral by "*ratio difficilis*" they establish with their source (Eco 1979, p. 220). Indeed, while symptoms and other types of traces proceed from a codified repertoire and are therefore easily recognisable (and falsifiable), it is more difficult to trace the imprint back to the printing agent. It seems to me defensible that a similar difficulty is encountered in recognising the type of vibration from its manifestation signs. In addition, it is noteworthy that in the case of vibration this recognition may result from a *perceptual semiotics*, embracing thus the idea, proposed by Jean Petitot, of a naturalisation of the perceptual structures (cf. Petitot et.al 1999; Petitot 2011), and overcoming Eco's long-lasting hesitations as regards the inclusion of perceptual operations in semiosis (cf. Eco 1997, pp. 106–107 and 124–125). Perceptual structures are apprehended in and through the observation of forms. It is thanks to them that we have the intuition of an intelligibility governing the morphology of the sensitive, that we experience a plastic thinking at work, and, in particular, that we perceive a vibration traversing forms.

There is, however, another aspect that can, in my opinion, bridge the gap between vibration and the production of imprints. Umberto Eco distinguishes two processes in reference to imprint: firstly, a "process of metaphor (it looks 'similar' to the printing agent, or it represents it)"; secondly, a process "of metonymy (one recognises the proof of a prior relationship of contiguity between the imprint and the printing agent)" (Eco 1979, p. 73, own translation). These two processes belong to two different semiotic processes: one iconic, the other indexical, and we can appreciate, throughout the argumentation, the caution the author displays when dealing with them.[9] It is this ambiguity between icon and index that, in his anthropologic approach of images, Georges Didi-Huberman does not hesitate to short-circuit as from the title of his book *La ressemblance par contact*,[10] whereas for Charles Sanders Peirce these were two different semiotic regards to a given object.[11] Still, this ambiguity seems to

[9]The criterion of resemblance is dependent on subjective, even though shareable, values (the imprint "looks 'similar'"), whereas production by physical contiguity is in general a supposition whose truthfulness must be verified ("one recognises the proof of a prior relationship of contiguity"). Eco (1979, p. 73, own translation).

[10]Georges Didi-Huberman refers to Peirce's "now classic distinction", but he doesn't dwell on explaining, from a theoretical point of view, the complicity ("*connivence*" in French) he spots between the "*index* (contact)" and the "*icon* (resemblance)" Didi-Huberman (2008, p. 45, own translation).

[11]Recognising the difficulty, if not the impossibility, of citing a case of pure index or finding a sign absolutely devoid of indexical quality, Peirce nevertheless stated: "Psychologically, the action of indices depends upon association by contiguity, and not upon association by resemblance or upon intellectual operations." And then: "[An index is a] sign, or representation, which refers to its object

reveal a constitutive doubleness not only of imprint but also of vibration, despite some differences.

Let us delve deeper in what sense it is possible to draw this parallel. First of all, the fact that the creation of forms through vibration does not necessarily proceed by direct and immediate contact, like for imprint, does not preclude that an iconic and indexical relation can be established, as evidenced in traditional photography generated by recording light vibrations on photosensitive paper. It is, of course, always a question of a relation that presupposes an act of reference and, as such, involves the problem of verifying its accuracy (cf. Eco 1992, p. 73). Afterwards, the fact that a force rather than an object, or a part of body, acts in *moulding* the arising morphology does not differentiate one creative modality from the other, since in imprint the source is in general presupposed and, in most cases, this presumption of physical connection with the printing agent can be empirically verified only by means of interpretation. In this regard, Didi-Huberman emphasised the worth of conceiving the notion of origin as an "eddy", according to the following definition given by Walter Benjamin in *The Origin of the German Tragic Drama (1928)*:

> Origin [*Ursprung*], although an entirely historical category, has, nevertheless, nothing to do with genesis [*Entstehung*]. The term origin is not intended to describe the process by which the existent came into being, but rather to describe that which emerges from the process of becoming and disappearance. Origin is an eddy in the stream of becoming, and in its current it swallows the material involved in the process of genesis. That which is original is never revealed in the naked and manifest existence of the factual [...]. (Benjamin 1928, English transl.: p. 45; quoted also in (Benjamin 1928, English transl.: p. 45; quoted also in Didi-Huberman 2008, note 12 p. 17)

The origin, which "needs to be recognized" and is "because of this, [...] something imperfect and incomplete" (Benjamin, ibid.), is therefore as dynamic in the imprint, analysed by Didi-Huberman, as in the vibration to which Eisenman subjugates his architectures. In one case as in the other, forms and forces at the origin of the work are reconstructed in a semiotic understanding of the analysed object, since the indexical reference to their origin is essentially built by conjecture and demonstration.

As for the iconic dimension of forms generated by imprint and vibration, both modes of production involve the question of similarity, but here differences are consequential. In imprint, the resemblance concerns the relation, generally *in a negative mode*, between the printing agent and the produced object or, also, the different specimens created from the same matrix or mould. According to Didi-Huberman, "imprint *duplicates*", it creates a double and, through repetition, generates "a *discrepancy*, albeit infra-slim [*inframince*], by which the produced *resemblances* [...] become [...] *unassimilable*" (Didi-Huberman 2008, pp. 230 and 302, own trans-

---

not so much because of any similarity or analogy with it, nor because it is associated with general characters which that object happens to possess, as because it is in dynamical (including spatial) connection both with the individual object, on the one hand, and with the senses or the memory of the person for whom it serves as a sign, on the other." Peirce (1931–1958, respectively, 2.306 and 2.305)

lation). In an apparently similar manner, in vibration forms that are alike,[12] not identical, are moulded following the frequency of oscillation they reproduce; however, they spatially diffract the latter in an inevitable dispersion, with resistances and interferences. Given their proximity and their specificities, we could hardly think of vibration as a particular case of imprint, since the imprint exists "only as singular" (Didi-Huberman 2008, p. 11, own translation), that is, in a one-to-one relation with the printing agent and in one and only temporality of creation for each specimen, whereas the vibration leads to proliferation in multiple forms, in a temporality that spatially extends in propagation.

There is a last distinction to be raised, perhaps the most important from a theoretical point of view; it concerns the parts affected by moulding or printing, on the one hand, and by frequency modulation on the other. For vibration does not only have an impact on the phenomenal appearance of the object, it does not only modify the outward or visible aspect of the matter set in oscillation, but it gives a structural shake which can alter and even destroy its architectonic configuration. Yet, until the signs of vibration do not disappear and vibration does not shatter the material consistency of the object supporting it, vibration tends to continue beyond the physic and spatial limits of its place of action. To that extent, it is not surprising to see it at work in the projects and achievements of Eisenman, for whom architecture is in becoming (cf. Eisenman 1997, new edition: p. 57). In contrast to the creating modality studied by Didi-Huberman, in which "*the shape obtained through imprint draws resemblance towards the death of art*" (Didi-Huberman 2008, p. 119, own translation), the morphodynamism resulting from vibration manifests the vitality of forces traversing the work, awakening an effect of presence, which indeed affects the perceiver.[13] The latter point relaunches the idea of vibration as transmission of a vital force which, from the work of art, ultimately reaches the beholder, who thus takes part in an energy flow. The passage from the morphology of the object to the perception of a structural dynamism and an expression of life induces to trace back from the work of art to its creative and receptive context. This lends to an undoubtedly romantic per-

---

[12]Following G. Deleuze and F. Guattari's *Mille plateaux* (*A Thousand Plateaus*), Eisenman distinguishes "self-similar repetition" from "selfsame repetition", the former being created by a "machinic" process that is "neither mechanical nor organic" (Eisenman 1997, new edition: p. 70).

[13]Jean-Paul Sartre observes that some images have the power to "make present" their object, be it absent, non-existent, or whose status of existence or inexistence would not be contemplated. He gives the example of various forms of portraiture (a photograph, a caricature, and a representation or mental image); however, his reasoning may expand to other kinds of signs establishing a psychophysical reference to an object. *Making present* is more than an indexical relation to the *having-being-there* that Roland Barthes will discuss in relation to photography. It concerns less the genesis than it mobilise a whole imaginary in the perceiver, thus participating in the definition of the object. Cf. Sartre (1940, in particular English transl.: p. 19). Instead of a *presentificatory* act, I would prefer to talk in terms of effects of presence, so as not to dull the semiotic character of this implication of the subject. Perhaps it is not useless to mention that, according to Jean-François Lyotard, the presence involves a focus on the act of perception and the temporality of the sensitive subject in the aesthetic moment: "Presence is the instant which interrupts the chaos of history and recalls, or simply calls out that «there is», even before that which is has any signification" (Lyotard 1998, new edition: p. 441). For a semiotic survey thereof, cf. Parret (2006).

spective, which, however, reveals the modalities of engendering and functioning of the emotional dimension conveyed by morphodynamics. This passage, involving the subject's sensitisation but also a metasemiotic stepping back—whence the presence effect—corresponds to the last of the three artistic phases called by Paul Klee "pre-creation, creation, post-creation", that is, respectively, the movement pre-existing the work of art, the active and creative movement directed on it and, finally, "the communication of the work's mobility to other, the beholders of the work" (Klee 1956, English transl.: p. 169). Soliciting the coenestesia of the subject's body,[14] vibration thus perhaps approximates to the English notion, derived from the same Latin origin, of *vibrancy*: the vital being, the being full of energy, and its power of irradiation. Whether the vibration is euphoric or destructive, it affects us deeply, "down to our subtlest parts", as Klee (1970, English transl.: p. 63) wrote or, more recently, as the string theory, evoked by Eisenman (cf. 2001, new edition: p. 124), proposes us to think.

References

Benjamin, Walter: Ursprung des deutschen Trauerspiels (1928) [English transl.: The Origin of German Tragic Drama. Its "Epistemo-Critical Prologue", pp. 27–56. Verso, London, New York. Also at http://cscs.res.in/dataarchive/textfiles/textfile.2009-12-28.3091501878/file (1998)]

Caliandro, Stefania: La vibration comme force morphogénétique de création. Ou le dynamisme architectonique dans l'œuvre de Peter Eisenman. In: Cygne noir. Revue d'exploration sémiotique, 1, 14 pp., Montreal. http://www.revuecygnenoir.org/numero/article/la-vibration-chez-eisenman (2013)

De Landa, Manuel: Deleuze, diagrams, and the genesis of form. ANY. Architect. N. Y., 23: Diagram work: data mechanics for a topological age, Jun, pp. 30–34 (1998)

Deleuze, Gilles: Francis Bacon. Logique de la sensation. Seuil, Paris (2002) [English transl.: Francis Bacon: The Logic of Sensation. Continuum, London, New York (2003)]

Didi-Huberman, Georges: La ressemblance par contact. Archéologie, anachronisme et modernité de l'empreinte. Minuit, Paris (2008)

Eco, Umberto: Segno. Isedi, Milan (1972) [French transl.: Le signe. Labor, Brussels (1988)]

Eco, Umberto: Trattato di semiotica generale. Bompiani, Milan (1975). New edition (1991)

Eco, Umberto: A Theory of Semiotics. Indiana University Press, Bloomington (1979)

Eco, Umberto: La production des signes. Librairie Générale Française, Paris (1992) [Revised excerpt from his Trattato di semiotica generale]

Eco, Umberto: Kant e l'ornitorinco. Bompiani R.C.S., Milan (1997)

Eisenman, Peter: Post/El Cards. A reply to Jacques Derrida. Assemblage, 12, Aug, pp. 14–17 (1990). New edition in: Eisenman P.: Written into the Void. Selected Writings 1990–2004, pp. 1–5. Yale University Press, New Haven and London (2007)

Eisenman, Peter: The Affects of Singularity. Architectural Design, 62, Nov-Dec, p. 51 (1992). New edition in: Eisenman, P.: Written into the Void. Selected Writings 1990–2004, pp. 19–24. Yale University Press, New Haven, London (2007)

Eisenman, Peter: Written into the Void. In: Eisenman, P. (ed.) Written into the Void. Selected Writings 1990–2004, pp. 79–86. Yale University Press, New Haven, London ([1994] 2007)

[14]On several occasions, Eisenman emphasises the centrality of the body in the sensitive and intelligible apprehension of his works, e.g. "Writing architecture cannot be thought without the already given physical experience of the body" (Eisenman [1994], 2007, p. 86).

Eisenman, Peter: Processes of the Interstitial: Notes on Zaera-Polo's Idea of Machinic. In: El Croquis: Peter Eisenman 1990–1997, 83, pp. 21–35 (1997) (bilingual edition in Spanish and English). New edition in: Eisenman, P.: Written into the Void. Selected Writings 1990–2004, pp. 50–71. Yale University Press New Haven, London (2007)

Eisenman, Peter: Diagram: An Original Scene of Writing. ANY. Archit. N. Y. 23: Diagram work: Data Mechanics for a Topological Age, June, pp. 27–29 (1998). New edition in: Eisenman, P.: Written into the Void. Selected Writings 1990–2004, pp. 87–94. Yale University Press, New Haven, London (2007)

Eisenman, Peter: Diagrams of Exteriority. In: Diagrams Diaries, pp. 164–209. Universe Publishing, New York (1999)

Eisenman, Peter: Autonomy and the Will to the Critical. Chien chu 42, Nov, pp. 58–63 (2000). New edition in: Eisenman, P.: Written into the Void. Selected Writings 1990–2004, pp. 95–99. Yale University Press, New Haven and London (2007)

Eisenman, Peter: A Matrix in the Jungle. In: Barzon, F. (ed.) The Charter of Zurich, pp. 28–37. Birkhäuser, Basel (2001). New edition in: Eisenman, P.: Written into the Void. Selected Writings 1990–2004, pp. 120–125. Yale University Press, New Haven, London (2007)

Eisenman, Peter: Blurred zones. In: Blurred Zones: Investigations of the Interstitial, Eisenman Architects 1988–1998, pp. 6–9. Monacelli Press, New York (2003). New edition in: Eisenman, P.: Written into the Void. Selected Writings 1990–2004, pp. 108–112. Yale University Press, New Haven, London (2007)

Eisenman, Peter: Digital Scrambler: From Index to Codex. Prospecta 35, pp. 40–53 (2004). New edition in: Eisenman, P.: Written into the Void. Selected Writings 1990–2004, pp. 133–150. Yale University Press, New Haven, London (2007)

Klee, Paul: Das bildnerische Denken, collected and edited by Spiller, J. Schwabe & Co., Verlag, Basel, Stuttgart (1956) [English transl.: Notebooks. The Thinking Eye, vol. 1. Lund Humphries, London (1961)]

Klee, Paul: Unendliche Naturgeschichte, collected and edited by Spiller, J. Schwabe & Co., Verlag, Basel, Stuttgart (1970) [English transl.: Notebooks. The Nature of Nature, vol. 2. Lund Humphries, London (1973)]

Lyotard, Jean-François: Les Transformateurs Duchamp. Galilée, Paris (1977). New bilingual edition: Les Transformateurs Duchamp. Duchamp's TRANS/formers. Van Looveren & Princen, Bruxelles (2010)

Lyotard, Jean-François: Discours, Figure. Klincksieck, Paris (1985)

Lyotard, Jean-François: L'instant, Newman. In: L'inhumain. Causeries sur le temps, pp. 89–99. Galilée, Paris (1988). New bilingual edition: L'instant, Newman (Newman: the instant). In: Parret, H. (ed.) Textes dispersés II: artistes contemporains (Miscellaneous Texts II: Contemporary Artists), pp. 424–442 and 425–443. Leuven University Press, Leuven (2012)

Parret, Herman: Épiphanies de la Présence. Essais sémio-esthétiques. Presses Universitaires de Limoges, Limoges (2006)

Peirce, Charles Sanders: Collected Papers of Charles Sanders Peirce. In: Hartshone, C., Weiss, P., Burks, A. (eds.), 8 volumes. Harvard University Press, Cambrige, MA (1931–1958) (As a matter of convention, references to this publication give the volume number, followed by a full stop, and the indication of the relevant paragraph number)

Petitot, Jean: La généalogie morphologique du structuralisme. In: Critique. Revue générale des publications françaises et étrangères, LV, 620–621: Claude Lévi-Strauss, pp. 97–122 (1999)

Petitot, Jean: Morphologie et Esthétique structurale: De Goethe à Lévi-Strauss. In: La Part de l'Œil, 25–26: L'art et la fonction symbolique, pp. 220–235. Bruxelles (2010–2011)

Petitot, Jean: Paper in answer to Jean-Claude Coquet. In: Nouveaux Actes Sémiotiques, 114, "Le Débat". Université de Limoges, published on 31 Jan 2011. http://revues.unilim.fr/nas/document.php?id=3641 (2011)

Petitot, Jean; Varela, Francisco J.; Pachoud, Bernard; Roy, Jean-Michel (eds.): Naturalizing Phenomenology. Issues in Contemporary Phenomenology and Cognitive Science. Leland Stanford Junior University Press, Stanford (CA) (1999)

Rowe, Colin; Slutzky, Robert:. Transparency: literal and phenomenal. Perspecta, 8, pp. 45–54 ([1955–56], 1963)
Sartre, Jean-Paul: L'imaginaire. Psychologie phénoménologique de l'imagination. Gallimard, Paris (1940). New edition: 1986. English transl.: The Imaginary. A Phenomenological Psychology of the Imagination. Routledge, New York, London (2004)

Chapter 4
Empathy, Meaning and Abstract Art

In aesthetics as in semiotics, renewed attention has recently been drawn to the phenomenon of empathy. This notion poses interesting questions to the definition of the relation that connects the beholder to the object of contemplation. The establishment of an empathetic contact influences indeed not only the subject's emotional state, but also the interpretative path the subject may develop. The study of the affective and/or cognitive components that might come into play currently differentiates approaches in the psychic field. Instead of concerning only psychology and psychoanalysis, this theme is relevant also in aesthetics, wherein the term indeed appeared for the first time in the late nineteenth century. By an historical reconstruction of its various meanings, I propose to pay regard to the existence of different conceptions of empathy, whose wild assimilation might further increase its misunderstanding, with the prejudices that ensue. Without wishing to aim at an empathetic method, this theoretical object will nevertheless be analysed in relation to the emergence of meaning in perception. Its mode of aesthetic apprehension will be considered as a tool for sense in art. Particular focus will be given to the historically attested relation between empathy and the artistic forms that are abstract from any figuration.[1]

For a Definition of Empathy

The notion of empathy is generally used in reference to a kind of non-verbal understanding that sets in between a subject and the intended object. This strange, elusive ability appears to be imperceptibly in action in interrelationships. It seems to evoke the sixth sense that leads to apprehend inner emotional states.[2] Although of an entirely

[1]The content of this essay has originally been presented under the title "Empathie et réception esthétique" at the 5th Congress of the International Association for Visual Semiotics organised in Siena, on 24 June 1998. This chapter is a translated re-edition of the French article (Caliandro 1999) partially resulting from the paper.

[2]Cf. Katz's definition (Katz 1963, pp. 1–11).

S. Caliandro, *Morphodynamics in Aesthetics*, Lecture Notes in Morphogenesis,
https://doi.org/10.1007/978-3-030-29631-5_4

mental order, empathic contact is mainly generated during a direct sensory activation. In addition, it may sometimes include a physical somatisation of the attitude or posture of the other. In relationships with other people for instance, one often becomes empathetic. One places themselves 'in another's position', identifies with the situation of the other and projects themselves by imagination out of themselves into the other. Without wanting to, it also happens to take the posture of the other, to appropriate the feelings of the other, to undergo a contagious influx of the other's euphoria, to feel physically concerned by the harms that seem to affect the other. Empathy would then be either mental or physical, or both. The subject finds itself involved in an 'as if' attitude: as if one were the other, as if one lived the other's feelings and sensations. When the 'as if' attitude is no more conscious, the subject temporarily forgets itself and abandons itself to merge into the other. This relationship would establish mostly with people to whom one feels close, but it is not uncommon to have an empathic contact with a stranger, an unknown person or even with an imaginary personage produced, for example, by literary fiction.

This very general definition actually brings together several different meanings that have been assigned to this word by various approaches. Some focus on the physico-biological factor, others on the substantially psychic value; some insist on the strictly affective and emotional relevance of empathy, while others include it in the cognitive framework. When I first looked at delving into this theme, I thought it had fallen into that historical oblivion that often covers notions of the past. Not only did I find a rich recent bibliography on this notion, but ironically the inquiry led me to a text of Kenneth Clark in which he denounced neglect of the theme, later refuted by his own research and by the response from other psychologists (cf. Clark 1980; Gladstein 1984, p. 38[3]). Empathy indeed seems to have been well studied in the psychological field with regard to intersubjective relationships. Switch of the roles in communication, contact with the client, psychotherapeutic function in medical treatment, transfer and countertransfer between patient and analyst, mother's relationship with the baby, learning and development of the child have all been examined in connection with empathy.[4] The research on the psyche seems thus to expand on this phenomenon and to envisage it in view of its effective practice.

According to Charette and La Flèche (1980), studies were conducted in three different orientations, which influence the results and prevent a univocal definition of the notion. Their classification highlights how a purely emotional value and/or a contribution to knowledge can be attributed to empathy. The first orientation underlines the affective reaction of the beholder. If a certain cognitive component is recognised to act, namely an understanding of the emotional experiences that are observed, empathy primarily requires "an emotional experience of the beholder, that is, a vicarious sharing of an affective experience" (ibid., p. 9, own translation). This first orientation is not so much concerned with the communication of the feelings and thereby with

[3]Cf. also Gerald A. Gladstein's article: "Empathy: An Unneglected Topic in Psychological Research" (1981).

[4]The bibliography being multifarious, I refer to the essay edited by Bohart and Greenberg (1997), which provides a general framework of research.

the relational dimension of empathy. It groups together the psychoanalytic theories of identification processes and the experiential approach interested in experimental study (cf. ibid., p. 20).[5] In contrast, the second orientation emphasises "the cognitive component of empathy that allows one person to understand the other" (ibid., own translation). For this cognitive orientation, empathy does not imply an emotional link. Phenomena of social understanding and of interpersonal perception are focused, or the relationship with the client. The client-centred approach also includes the therapeutic relationship insofar as the therapist's understanding is not transformed into a loss of identity and an identification process. The fundamental character indeed lies in the communicative aspect determined by empathy (cf. ibid., pp. 20–22). Finally, the third integrative orientation seeks "to include both cognitive and affective components of empathy" in one conceptual model (ibid., p. 22, own translation). Despite some internal contradictions, the approaches that fit into the latter orientation aim to reconcile the two previous orientations. "The affective component, which consists of feeling themselves like the other", would then be "inseparable from the cognitive process because these two components work in a complex and interactive way" (ibid, p. 23, own translation).

Apparently, the notion of empathy has not been neglected in the contemporary intellectual panorama. The development of a wide variety of research and different approaches seems, on the contrary, to testify to the topicality of the theme. However, if we look back at the definition of empathy Clark quotes from the *Webster* dictionary, we perceive a lasting oversight. This dictionary actually has two main entries. The first describes empathy as an imaginary projection of a subjective state into an object, which consequently appears to be impregnated with it. It is no coincidence that Clark instead favours the relationship between subjects in spite of the possible relationship between subject and object. For him, as for other psychologists, empathy corresponds rather to the second definition in the dictionary centred on interpersonal relationship. He writes:

> In the unabridged *Webster's Third New International Dictionary*, empathy is defined as "1: the imaginative projection of a subjective state whether affective, conative, or cognitive into an object so that the object appears to be infused by it… 2: the capacity for participating in or a vicarious experiencing of another's feelings, volition, or ideas and sometimes another's movements to the point of executing bodily movements resembling his".
>
> It is of some significance that the second dictionary definition is the one that I would have placed as the first and the most relevant (Clark 1980, p. 188).

The idea that a subject can merge with an object seems in fact to have been forgotten for a long time and only recent research in the twenty-first century has rehabilitated it. However, if we take a closer look, this definition has existed since the very origin of the word. In the wake of Schopenhauer's idea that, in the contemplation of art, the

---

[5]The authors of the classification note that, even if the psychoanalytic model of empathy is affective, the therapeutic process does not imply an empathetic involvement of the analyst with the patient. Quite the opposite, a cognitive detachment is asked of the therapist (cf. Charette and La Flèche 1980, p. 20).

subject completely loses itself in the object,[6] the notion of empathy was first created in the late nineteenth century to account for aesthetic phenomena. From the outset, it did not concern only psychology but aesthetic reception of works of art. I propose then to reconsider this notion according to its first historical meanings, trying to distinguish it from the transformations of sense it has known thereafter. I will finally look at the contributions it can give to a semiotics of art and whether its modes of interpretation can currently be judged founded for an aesthetic reception.[7] Instead of following the simple opposition between affection and cognition or looking for a possible third integrative way, as in the case of empathy between subjects, I will try to point out that this notion touches fundamental questions of aesthetics and that it has not yet been fully comprehended. By returning to the texts of its earlier theoreticians, I will try in particular to analyse some aspects that make it possible to understand empathy as a tool of meaning in art.

Empathy in Art

Whilst the concept of empathy seems to go back to Aristotle,[8] the German term *Einfühlung* was coined for the first time in 1873 by the son of Friedrich Theodor Vischer, Robert Vischer, then taken up by the father in his aesthetic reflections.[9] This term, which will be translated in English as *empathy* by Edward Titchener in 1909,[10] differs from the term *Mitfühlung* (sympathy), since it does not indicate a simple sharing (feeling with) of an experience, but to perceive it as one were in unity, a feeling in unison. On the other hand, *Einfühlung* is also different from *Zweifühlung* (dialogue), wherein both parties keep their identity separate (cf. Barnes and Thagard 1997, p. 707; Katz 1963, p. 82; and Gladstein 1984, pp. 43–45).

Furthermore, a very accurate terminological choice seems to have determined the actual coinage of the word, since in his essay Robert Vischer differentiates *Einfühlung* in particular from *Anfühlung* (translated in English as "attentive feeling"), *Zufühlung* ("immediate feeling") and *Nachfühlung* ("responsive feeling") (Vischer 1873, English transl.: pp. 106–107; cf. also Mallgrave and Ikonomou 1993, p. 22). These terms in fact designate processes that go from the exterior of the object (its

[6]Cf. also Morgan (1996, p. 320), quoting from the English translation: Arthur Schopenhauer. *The World as Will and Representation*. 2 vol. Col., Indian Hills. 1958, pp. 178–179.

[7]In a 1992 article, Crozier and Greenhalgh proposed updating the notion of empathy for a psychological theory of responses to works of art. According to the authors, the "Empathy Principle" would make it possible to go beyond the limits of relativism and formalism in art criticism (Crozier and Greenhalgh 1992, p. 86).

[8]In *Rhetoric*, III, 2, concerning the problem of metaphor, Aristotle deals with the projection of sensations, emotions and human attitudes in inanimate objects (cf. Gilbert and Kuhn 1954, p. 537).

[9]In particular, see: Friedrich Theodor Vischer. *Kritische Gänge*. Stuttgart: Verlag der J. G. Cotta. 1873. Cf. Morgan (1992, p. 233); Gladstein (1984, pp. 39–40); and Gilbert and Kuhn (1939, p. 506).

[10]Cf. Edward B. Titchener, *Lectures on the Experimental Psychology of the Thought-Process*, Macmillan, New York, 1909, p. 21, quoted in Gladstein (1984, p. 40), also Rader (1960, p. 367).

shape) to its centre, but which may not reach the intrinsic quality of the object. According to this theoretician, empathy leads to retrace the object and its external form from its core (cf. Vischer 1873, English transl.: p. 108). It deploys a projection and an incorporation of the physical form of the subject in the form of the object. It enables the animation of the plant world, the anthropomorphism of animals or to feel another human being as a double of oneself (cf. ibid., English transl.: pp. 104 and 106). Taking up his father's theories on the symbolism of form, Robert Vischer did not conceive, however, of empathy as a road to mere pantheism, which he considered the final stage of animation of nature (cf. ibid., English transl.: pp. 109–112), but he understood it as a means by which one charges any form, even inorganic phenomena, with emotional content. Thus, he associated this notion also with the motion effects of static forms such as vertical, horizontal or curved lines and planes (cf. ibid., English transl.: pp. 89–90). He, hence, related it to the modes of optical perception as well as to the artist's work of aesthetic production (cf. ibid., English transl.: pp. 103 and 112ff.).

From the beginning, however, the notion has taken on various meanings, and the expression coined by Robert Vischer will sometimes acquire opposing significations. On the one hand, it has been understood as an experience of fusion abolishing any barrier between subjective feeling and objective reality, often in a pantheistic view. This is, for instance, Johannes Volket's conception, resumed however in the doctoral thesis of his student Heinrich Wölfflin, who freed it from all pantheism.[11] Such an idea of a total immersion in nature continues today when we use the term of empathy in a romantic sense or it is diminished as scientifically unmanageable. On the other hand, these debates have sketched a reflection on the emotional reactions to optical phenomena, which seems to anticipate current study in psychology of vision. In this sense, some theorists of that time used the notion of empathy to put forward a theory of perception according to which the human feeling projects itself into visual forms through the interpretative action of the eye (cf. Morgan 1992, p. 235; 1996, p. 321). Without reducing the subjective value, they sought to define the sensations that sensory stimulation may generate in a state of mind. They focused, therefore, on the most instinctive and primordial stages of significance, that is, anterior to any extensive deployment in what they called the association of ideas.

These thoughts on empathy concerned, at least from a theoretical point of view, painting, sculpture as well as, for many references, the musical field. It was up to Wölfflin to extend them to applied arts and, in particular, to architecture. Without ever discussing it directly, but making use of substitutive terms such as sympathy, the young Wölfflin seems to hold a position apart from the theoreticians of empathy. The purpose of his thesis indeed was to analyse how architectural forms are likely to express emotions, that is, how the effect or impression that one receives from them is understood as an expression of the object itself (cf. Wölfflin 1886, English trans.: pp. 149–150). His approach rejects both the pantheism of Volket and Friedrich

[11] Applying the theory of empathy to architectural forms, Wölfflin (1886) argued that instinctively we animate everything. He explained this animation of the external world as a subordination of all appearances to the image of our body.

Theodor Vischer and the physiological hypothesis that expression is determined by the kinaesthetic movement of the eye. He nevertheless takes up Volket's theory to put forward the idea of the centrality of the body in experiencing sensations. It is the body as a whole that experiences and makes meaningful any physical form. The projection that animates the objects would, therefore, not be the result of a mental process based on imagination, as suggested by Robert Vischer and Lotze, but arises from the direct involvement of sensory organs (cf. ibid., pp. 150–158).

The importance that the first Wölfflin accorded to the body and its physical senses will still be a key topic of the discussions on empathy that continue in the late nineteenth century. In her first essay for instance, the Englishwoman Violet Paget writing under the pseudonym Vernon Lee emphasises the relevance that the posture of the beholder acquires during the aesthetic reception: according to her, only the imitation on the part of the subject would give the full empathic appreciation of the contemplated object.[12] Afterwards, in the wake of Theodor Lipps's writings, she will however review her affirmations, and the attitude to imitate ("mimicry") will then become a process more imaginary than physical (cf. Lee 1913, new edition: pp. 372–373). Returning to Wölfflin and his doctoral thesis, the most striking feature concerning his psychology of architecture is that the special attention he draws to the bodily perspective of the receiver will considerably contrast with his subsequent achievements. In the following studies on structural modes of visualising form, Wölfflin seems indeed to have completely set aside and gone beyond this early research. Yet, conceptual connections could be established concerning this shift from the psychology of perception to questions on form and explain thus the influence of pure visibility on contemporary theories. These connections should be explored further, and this is not the right place, as they might uncover unexpected facets of historical formalism.

Empathy and Abstraction

At the turn of the twentieth century, the notion of empathy lent itself to new developments. The growing favour that this topic was going to assume was due to the resonance of Theodor Lipps's writings, which mainly contribute to disseminate the

[12]Cf. Gladstein (1984, pp. 39–42) and Rader (1960, pp. 368–369). Lee's first article on empathy, "Beauty and Ugliness", written under the influence of her friend Karl Groos, a philosopher of empathy, appeared in the *Contemporary Review* in 1899, two years before she became acquainted with the essay Lipps published in 1897.

notion.[13] This will subsequently lead to an accustomed relationship with contemporary artistic research. The experimentation of nonfigurative modes of expression on the part of the artists suggests a work on vision and its perceptive effects, which may be entangled with the reflection on empathy. The proximity of this notion to that of abstraction, however, turns out to be critical. The arguments of the beginning of the century are actually problematic in regard to the historical tradition. By confronting Lipps's theoretical position with the succeeding developments, we perceive that not only does empathy become the fulcrum for other understandings, but that certain contributions deflect the meaning of the notion from its original sense.

Resuming the debate over the value of body and senses in the empathic contact, Lipps argued that the sensual appearance of a work of art may be the object of aesthetic pleasure but not its cause, which resides rather in the ego of the beholder. The objectification of the ego in the work of art and the sensory perception in the beholder constitute indissolubly the aesthetic pleasure. Empathy would be precisely that relationship by which "the object is myself and by the very same token this self of mine is the object". Empathy is thus "the fact that the antithesis between myself and the object disappears, or rather does not yet exist" (Lipps 1903, English transl. (1979): p. 372). By opposing the theory of 'elements' and freeing itself from any metaphorical sense, empathy emphasised the importance of the subject in the perception. To the anthropomorphism of the vision corresponded also a modification of the beholder who, relieved of their physical presence and emptied of their bodily sensations, would experience the joy of engaging totally in contemplation. This modification of the subject could finally manifest through an involuntary impulse to imitate the attitude of the object, but for Lipps this was only a symptom, not the cause, of the aesthetic pleasure (cf. ibid., English transl. (1979): pp. 375–377).

In his doctoral dissertation entitled *Abstraktion und Einfühlung*, Wilhelm Worringer re-elaborated the ideas of Lipps by presenting the most famous application of empathy in art history. His interpretation, however, moves away considerably from Lipps's conception. For Worringer, in fact, empathy is only one of the two poles of the human attitude towards art, the other being abstraction (cf. Worringer 1908, English transl.: p. 4). In addition, empathy is associated with a search for organic effects in art that leads to naturalism, while abstraction by its inorganic forms leads to

[13]The fundamental texts of Lipps are *Raumästhetik und geometrisch-optische Täuschungen*, Schriften der Gesellschaft für psychologische Forschung, IX–X (Johann Ambrosius Barth, Leipzig, 1897); *Ästhetik. Psychologie des Schönen und der Kunst*, first part, *Grundlegung der Ästhetik* (Leopold Voss, Hamburg and Leipzig, 1903); and *Zur Einfühlung* (Englemann, Leipzig, 1913). Also Paul Stern published in 1898 *Einfühlung und Association in der neueren Ästhetik* (Leopold Voss, Leipzig):

> In it he criticized the use of the term "empathy" as metaphorical and therefore misleading, and argued for association as a psychological basis for aesthetic experience. […] Stern criticized such proponents of empathy as Vischer and Volket, who had linked empathy to pantheism (Morgan 1992, p. 237).

Other contemporary proponents of empathy are Karl Gross (*Einleitung in die Ästhetik*, 1892, and *Der ästhetische Genuss*, 1902), Victor Basch (*Essai critique sur l'esthétique de Kant*, 1896), Hermann Lotze and Wilhelm Wundt.

style (cf. ibid., English transl.: pp. 31–35 and 55–56). His terminology deserves some remarks, though. According to him, naturalism is not an imitation of nature which, in itself, has no claim to art. Moreover, his notion of abstraction must be understood as a tendency to transcendental spirituality, in opposition to the immanent anthropomorphism of empathy. For that reason, Worringer places Egyptian and Nordic art into abstraction, whilst classicism and Italian Renaissance would be empathic (cf. ibid., English transl.: pp. 128–130; cf. also Morgan 1992, p. 238).

Opposing empathy to abstraction, Worringer reversed all previous aesthetic reflection at once. Vischer had introduced empathy to account for the aesthetic significance of the most inorganic and abstract forms, thus going beyond the distinction that separated them from figurative painting. Empathy enabled to connect a symbolic content to the colour, the tone, the line and the surface which, in themselves, would not have reached the aesthetic. Vischer considered, for instance, landscape painting—which not incidentally he described as 'musical'—as a genre wherein the abstract phenomena or formal artistic elements are imbued with a disposition of the mind (cf. ibid., pp. 233–234). Thanks to this empathic contact with the work of art, the beholder would transpose his or her state of mind into the appearance of non-representational forms and charge them with a symbolic value. Similarly, in a first essay in 1897, Lipps had studied empathy in relation to geometric forms; only afterwards he extended the meaning to the perception of the figurative. In 1903, he still wrote that a colour per se, permeated with the empathic mood, acquires internal vitality (cf. ibid., p. 235). Finally, Vernon Lee had also noticed that lines and forms animate themselves in the act of perception. All this has nothing to do, therefore, with the inorganic dimension that, according to Worringer, would deprive abstract art of any sensory evocation. As Kandinsky inherited a few years later,[14] the empathic expressiveness was not reserved for naturalistic figuration, but operated even more in non-representational forms.

Reception and Meaning

Since the second decade of the twentieth century, the theory of empathy in art criticism has been intensely contrasted (cf. ibid., p. 237; Gilbert and Kuhn 1939, pp. 539–540). In that way was, among others, Arnheim's attack, representative of the psychology of perception. In his search for a structure of perception, expressiveness could not result, for him, from the empathy of the subject, but from the qualities intrinsic to the forms of the object (cf. Arnheim 1964, pp. 362–368; Rader 1979, p. 334)[15]. Apart from a few historical studies, general negligence has surrounded the notion up to the

[14] *Über das Geistige in der Kunst* by Wassily Kandinsky is dated 1912. Cf. also Morgan (1992, pp. 238–239). Kandinsky will also pursue the parallel with music by studying sound in relation to colour.

[15] In addition, Arnheim pursues the confusion between empathy and sympathy; cf. Arnheim (1964, p. 368); also Gladstein (1984, pp. 43–45).

end of the century. An essay, which did not actually have many echoes as far as I am aware, proposed, however, to revisit empathy as a tool for aesthetic interpretation (cf. Crozier and Greenhalgh 1992; also *supra* footnote n. 7). I would, therefore, like to reconsider the modes of such a reception, at least in its historical forms, and see whether the theory of empathy can provide insights into the semiotics of art in terms of cognition and/or affection and, in such case, how it can be a tool for meaning.

If we follow Charette and La Flèche's classification concerning empathy between subjects,[16] we would be tempted to put forward a critique of the traditional approaches to empathy in art that seem to correspond only to the affective orientation. The only cognitive contribution would in fact concern the emotional state of the receiver, and these theories do not care about the communicative aspect of elaborating and transmitting impressions. Everything is centred on the beholder's psychological feelings. As Freud explicitly stated, the significance that empathy has for the intellectual life is left on one side to "limit ourselves to the immediate emotional effects of identification".[17] The psychological dimension of empathy seems indeed to halt at the subjective perception, at least following Lipps's thought. Empathy, the latter writes, is a non-objective activity, which could of course get to be objective when contemplation becomes retrospective, i.e. by detaching from immediacy, but, he continues, this is not relevant to him:

> This activity is not objective. It is not anything that stands opposite to me. Just as I do not feel active over against the object, but *in* the objet, so I do not experience joy over against my activity, but in it. I feel happy or blessed *in* it.
>
> To be sure, my own activity may become objective to me namely, when it is no longer my present activity but when I contemplate it in retrospect. But then it is no longer immediately experienced, but only remembered in imagination. And thus it is objective. This imagined activity, or more generally this imagined self, can then also be the object of my joy. But this we are not discussing. We are now only concerned with the immediately experienced activity (Lipps 1903, English transl. (1979): p. 372).

Also, in his chromatic analysis each colour has its humoral character, but Lipps refuses to list them since they depend on the beholder's individuality (cf. Morgan 1992, p. 235). By linking to perceptual immediacy and personal variability, empathy does not reach in itself an intersubjective level. Furthermore, according to Katz's study, empathy between a subject and an object would be entirely projective (cf. Katz 1963, pp. 92–93). Therefore, the inevitable errors due to interpretative excesses (cf. ibid., pp. 163–164), instead of being limited through an explicit and verifiable communication, are paradoxically left to the free will of a possible narcissistic projection (cf. ibid., p. 169). Reception in terms of pure emotional meaning in art is thus likely to dip towards an extreme subjectivity. Even the most daring semiotic practices, such as that of Fernande Saint-Martin which intends to use psy-

[16] See, *supra*, their classification of the affective, cognitive and integrative orientations.

[17] Freud (1922, new edition: p. 66). Freud explains empathy mainly through the process of identification (cf. ibid., pp. 60–70, esp. 70). Apparently, Freud repeatedly referred to empathy (15 times in the corpus of his writings) without ever using it as a key instrument for his analysis (cf. Eagle and Wolitzky 1997, p. 222).

chological and psychoanalytic material to disclose art meaning, seem at odds with the perceptual immediacy that Lipps acknowledged to the empathic contact.[18]

However, if we take into consideration the interpretative development that empathy can engender, one must nevertheless recognise a cognitive value that poses empathy at the beginning of a process of signification. In a perspective attentive to the morphogenesis of meaning, such as that elaborated by Jean Petitot in the wake of René Thom, one could probably attribute to it a *semio-physical* dimension (cf. Petitot 1996, p. 68; also Petitot 1990). Empathy would then be not only a driving force that leads to the quest for meaning, but also seems to condition and somehow predetermine the results of such a quest. In order to become an interpretative tool, it should alternate, as Katz already suggested (cf. Katz 1963, p. 26), with an operation of detachment. This would allow the transition from subjective involvement to a critical attitude. Finally, the notion of empathy can be reconsidered according to Aby Warburg, who linked it to social and cultural memory. The *pathos* would then become an emotional charge, an energetic way for the transmission of works of art. For Warburg, biomorphic empathy and rational detachment are indeed two indissociable poles of the constant schizophrenia of humanity: by this way artistic images revive thanks to the interpretations given by posterity.[19]

To summarise, the main features of aesthetic empathy are the subjective implication preventing the distinction between subject and object, the perceptual immediacy and the non-communicative dimension. The latter brings empathy closer to the concept of "*phatic communion*" which, for Malinowski (1949, p. 315), refers to linguistic exchanges that do not contain informative communication.[20] Likewise, the disposition to "a total fusion of the subject and the object" is connoted by Algirdas-Julien Greimas as "[c]ognitively elusive" (Greimas 1987, pp. 73 and 72, own translation). However, a cognitive importance should be recognised, because empathic contact can trigger emotional and aesthetic interpretations. The notion of empathy finds a correspondence, perhaps, with Charles Sanders Peirce's notion of "feeling". For him, the "feeling", which is different from the sensation and the emotion,[21] is characterised by its *presentness*, namely the immediacy of a conative or perceptual attitude con-

[18]The first reactions in the contemplation of the work of art, when the perceptual relationship is superficially sketched, depend mainly, according to Fernande Saint-Martin, on the tonic equilibrium of the body (cf. Saint-Martin 1989, p. 124). "These reactions, often preliminary to an *adequate* understanding of the work of art, which can only be done over a long period of time, should no longer, in our opinion, be considered as «aesthetic» judgements, even though they form the basis on which surveys on aesthetic perception are carried out" (ibid., p. 125, own translation).

[19]Cf. Gombrich (1970, p. 253); Caliandro (1997–1998, pp. 89–90). For the influence of Friedrich Theodor Vischer's theories on Warburg, cf. Gombrich (1970, pp. 72ff).

[20]An excerpt from Malinowski's essay is quoted by Émile Benveniste in "L'appareil formel de l'énonciation"; cf. Benveniste (1974, pp. 87ff).

[21]The difference between feeling and emotion is always clear in the *Collected Papers of Charles Sanders Peirce*, while the terms "feeling" and "sensation" are sometimes indistinctly positioned in the "Firstness" of the cognitive path. Yet, in sensation a degree of consciousness would be added to the simple feeling (Peirce 1931–1958, 5.295 and 7.586). It is striking that, although the taxonomy is not superposable, Robert Vischer also distinguishes between sensation, feeling and emotion (Vischer 1873, English transl.: p. 109).

cerning an ego that does not yet recognise itself (Peirce 1931–1958, 1.307, and also 2.85 and 5.44):

> Introspection is wholly a matter of inference.[...] One is immediately conscious of his Feelings, no doubt; but not that they are feeling of an *ego*. The *self* is only inferred. There is no time in the Present for any inference at all, least of all for concerning that very instant. Consequently the present object must be an external object, if there be any objective reference in it. The attitude of the Present is either conative or perceptive. Supposing it to be perceptive, the perception must be immediately known as external – not indeed in the sense in which a hallucination is *not* external, but in the sense of being present regardless of the perceiver's will or wish. Now this kind of externality is conative externality. (Ibid., 5.462)

The "feeling" depends indeed on the bodily organisation and, on several occasions, Peirce defends the scientificity of an anthropomorphic conception (cf. ibid., 5.46-47, 5.539-540 and 7.586). Even without explicitly referring to empathy, Peirce, however, attacks contemporary German theories because they exaggerate this primary element, thus reducing logical reasoning to aesthetic satisfaction.[22] The strength of the "feeling" would be instead in the possibility of transforming itself into an emotional, an energetic and a logical interpretant (cf. ibid. 5.475-476) and then opening the meaning of art by an infinite semiosis.

To Conclude

Empathy would hence pertain to the *firstness*. Following Lee's examples, a line is said to extend from point A to B, whilst actually the line extends nowhere and that indeed is the effect of the gaze. The same goes for the upward effect of a mountain, which in fact does not rise except in human perception (cf. Lee 1913, new edition: pp. 370–371). Subject and object are thus united in an undifferentiated amalgam; they are not yet semiotised. They will be semiotic entities only in the ensuing interpretative development. As Peirce states, the subject's introspection is produced solely by inference. One is aware of his or her feelings, but not aware that the latter are the feelings of an ego. The self can only be inferred and, in the immediacy of the present, this inference is not possible (cf. Peirce 1931–1958, 5.462 cited *supra*). Nonetheless, according to him: "Feelings, in the sense in which alone they can be admitted as a great branch of mental phenomena, form the warp and woof of cognition"; and when the semiotic process becomes conscious, these phenomena generate cognitive "mental growth" (ibid., 1.381). We could say that, through empathy, the empirical subject introduces itself into the morphology of the perceived object by inscribing own physical and psychic dimensions. Then, thanks to the critical detachment, the passage to raising an awareness concerning the subject and the object will take place. Besides, in this

[22]Peirce criticises contemporary German theories in general (Peirce 1931–1958, 1.304 and 2.165), but a passage in which he criticises the notions of "Fusion, Assimilation, Synthesis, Complication, Apperception" put forward by the English associationism also suggests a possible reference to the theory of empathy (ibid., 7.377).

firstness, the perceived is recognised as an *image* in the sense of Peirce, i.e. for having *hypoiconic* value.

This gives a better understanding of why the main proponents of empathy, starting from Robert Vischer himself, theorised this notion by referring to the most abstract elements of the artistic representation. Not sharing the formalist viewpoint,[23] they discovered the subjectivity of vision, in particular, from nonfigurative forms, where the non-objectivity was more obvious, to build a theory of perception embracing all artistic phenomena.[24] In this sense, the idea of a *rhythmic impression of the form*, occurring even in still images, seems symptomatic of the introduction of the potentialities of the subject—in this case the time of perception—in the aesthetic contemplation (cf. Vischer 1873, English transl.: p. 97).

The aesthetic object cannot be apprehended by senses, nor even less become signifying if not by being incorporated into the interpreting subject. The latter reveals, therefore, to be the essential locus of meaning. As Fontanille also noted, the question of the emergence of meaning in perception compels us to "go beyond the definition of the sign as a reciprocal presupposition between the signifier and the signified, or between the expression plane and the content plane" (Fontanille 1999, p. 228, own translation). The work of art turns out to be meaningful only thanks to an interpreting system that takes the work of art as object and recognises in it significance, viz. a semiotic status. The recognition of a semiotic value really seems to be more problematic in the case of abstract art, where a predetermined significance can hardly be spotted. I would then like to propose that empathy works, in the specific case of abstract art as well as any aesthetic phenomenon, as a way for including it in the framework of an interpreting semiotics, that is to say, as a tool to let the object under analysis become signifying.

Considering that, according to Peirce's theory, meaning is supplied in all the phases of the process of interpretation (cf. Eco 1997, p. 99, referring to Nesher[25]), the recognition of hypoiconic forms, i.e. of an expression plane, could not occur without the constitution of a first core of sense from a cognitive point of view.[26] We should, therefore, observe from a new perspective at these first perceptive stages that are set at the genesis of meaning. In a 1970 essay concerning the reading of the painting,

---

[23]See the theory of *elements* or the thoughts of Herbart, in Morgan (1992, p. 232).

[24]Peirce himself has linked the notion of hypoicon to painting:

> If a substantive be wanted, an iconic representamen may be termed a *hypoicon*. Any material image, as a painting, is largely conventional in its mode of representation; but in itself, without legend or label it may be called a hypoicon. (Peirce 1931–1958, 2.276)

According to the way of firstness in which they participate, hypoicons are thereafter distinguished into *images*, *diagrams* and *metaphors* (cf. ibid., 2.277).

[25]In particular, Umberto Eco refers to the article of Dan Nesher: "Are There Grounds for Identifying «Ground» with «Interpretant» in Peirce's Pragmatic Theory of Meaning?", Transactions of Charles Sanders Peirce Society, XX, 3, summer, 1984, pp. 303–324.

[26]It will also be noted that in 1975 Hubert Damisch took up the notion of hypoicon to account in particular for the art of the twentieth century (cf. Damisch 1975, p. 11). An article written by Marie Carani (1998) also set hypoicon as a paradigm for visual arts, with special regard to abstraction.

Louis Marin, for example, refuted "the immediate which was only nonsense" and he delegated any signification to the mediation of the discourse (Marin 1970, new edition: p. 62, own translation). A few lines before, however, he was not so sure: "Sense: is there thus no sense in seeing? Yes, no doubt, but immediate and this is a kind of nonsense" (ibid., own translation). This "kind of nonsense", rather than simple "nonsense", would finally be a rudimentary, primitive sense, of which any successive interpretative deployment must nevertheless take account. It is noteworthy that this first core of meaning is built visually.

References

Arnheim, Rudolf: Art and Visual Perception. A Psychology of the Creative Eye. University of California Press: Berkeley, Los Angeles (1964)

Barnes, Allison; Thagard, Paul: Empathy and analogy. Dialogue. Can. Philos. Rev. /Rev. Canadienne de Philosophie, XXXVI, 4, autumn, pp. 705–720 (1997)

Benveniste, Émile: Problèmes de linguistique générale II. Gallimard, Paris (1974)

Bohart, Arthur C.; Greenberg, Leslie S. (eds.): Empathy Reconsidered. New Directions in Psychotherapy. American Psychological Association, Washington DC (1997)

Caliandro, Stefania: Nachleben de Warburg. Visio, 2, 3, autumn-winter, pp. 87–103, Canada (1997–1998)

Caliandro, Stefania: Empathie, signification et art abstrait. Visio, 4, 2, summer, pp. 48–57, Canada (1999)

Carani, Marie: L'hypoiconicité comme paradigme des arts visuels. Visio, 3, 1: Iconicité, hypoiconicité/hypoicons, spring, pp. 65–84, Canada (1998)

Charette, Réjan; La Flèche, Jo-Anne: Les composantes de l'empathie: état des recherches. Revue Québécoise de Psychologie, 1, 2, May, pp. 13–27 (1980)

Clark, Kenneth B.: Empathy. A neglected topic in psychological research. American Psychologist, 35, 2, Feb, pp. 187–190 (1980)

Crozier, W. Ray; Greenhalgh, Paul: Beyond relativism and formalism: The empathy principle?, Leonardo 25, 1, pp. 83–87 (1992)

Damisch, Hubert: Semiotics and iconography. In: Sebeok, A.T. (ed.). The Tell-Tale Sign: A Survey of Semiotics. The Peter Ridder Press, Lisse (Netherlands), offprint, pp. 27–36 (1975)

Eagle, Morris; Wolitzky, David L.: Empathy: a psychoanalytic perspective. In: Bohart, A.C.; Greenberg, L.S. (eds.). Empathy Reconsidered. New Directions in Psychotherapy, American Psychological Association, Washington DC, pp. 217–244 (1997)

Eco, Umberto: Kant e l'ornitorinco. Bompiani R.C.S, Milan (1997)

Fontanille, Jacques: Sémiotique et littérature. Essais de méthode. PUF, Paris (1999)

Freud, Sigmund: Group Psychology and the Analysis of the Ego. London, International Psycho-Analytic Press (1922). New edition. London, The Hogart Press (1949)

Gilbert, Katharine Everett; Kuhn, Helmut: A History of Esthetics. The Macmillan Company, New York (1939). 2nd revised and extended edition: Indiana University Press, Bloomington (1954)

Gladstein, Gerald A.: Empathy. An Unneglected Topic in Psychological Research. American Psychologist, 36, 2, pp. 224–225, Washington, DC (1981)

Gladstein, Gerald A.: The historical roots of contemporary empathy research. Journal of the History of Behavioral Science, 20, Jan, pp. 38–59 (1984)

Gombrich, Ernst H.: Aby Warburg. An Intellectual Biography. The Warburg Institute, University of London, London (1970). New edition (1986)

Greimas, Algirdas-Julien: De l'imperfection. Pierre Fanlac, Périgueux (1987)

Katz, Robert L.: Empathy. Its Nature and Uses. The Free Press of Glencoe Collier-Macmillan, London (1963)
Lee, Vernon (Pseudonym of Violet Paget): Empathy. The Beautiful. Cambridge (1913). New edition in: Rader, M. (ed.). A Modern Book of Esthetics. An Anthology, 3rd edition. Holt, Rinehart and Winston, New York, Chicago, San Francisco, Toronto, London, pp. 370–374 (1960)
Lipps, Théodor: Einfühlung, innere Nachahmung und Organempfindungen. Archiv für die gesamte Psychologie, vol. I (1903) [English trans.: Empathy, Inner Imitation, and Sense-Feelings. In: Rader, M. (ed). A Modern Book of Esthetics. An Anthology, 3rd edition. Holt, Rinehart and Winston, New York, Chicago, San Francisco, Toronto, London, pp. 374–382 (1960); 5th edition: Holt, Rinehart and Winston, New York, San Francisco, Dallas, Montreal, Toronto, London, Sidney, pp. 371–378 (1979)]
Malinowski, Bronisław: The Problem of Meaning in Primitive Languages. In: Ogden, C.K.; Richards, I.A. (eds.) The Meaning of Meaning. A Study of Language upon Thought and of The Science of Symbolism, with essays by B. Malinowski and F.G. Crookshank. Routledge & Kegan Paul Ltd, London, pp. 296-336 (1949)
Mallgrave, Harry Francis; Ikonomou, Eleftherios (eds.): Empathy, Form, and Space. Problems in German Aesthetics, 1873–1893. The Getty Center for History of and the Humanities, London (1993). New edition: Santa Monica (1994)
Marin, Louis: Textes en représentation. Critique, November (1970). New edition: Études sémiologiques. Écritures, Peintures. Klincksieck, Paris, pp. 61–83 (1971)
Morgan, David: The idea of abstraction in German theories of the ornament from Kant to Kandinsky. Journal of Aesthetics and Art Criticism, 50, 3, summer, pp. 231–242 (1992)
Morgan, David: The Enchantment of Art: Abstraction and Empathy from German Romanticism to Expressionism. Journal of the History of Ideas, 57, 2, April, pp. 317–341 (1996)
Peirce, Charles Sanders: Collected Papers of Charles Sanders Peirce. In: Hartshone, C., Weiss, P., Burks, A. (eds.), 8 volumes. Harvard University Press, Cambrige, MA (1931–1958). Read edition: 1960 (As a matter of convention, references to this publication give the volume number, followed by a full stop, and the indication of the relevant paragraph number)
Petitot, Jean: Le Physique, le Morphologique, le Symbolique: remarques sur la vision. Revue de synthèse, IV series, 1–2, Jan–Jun, pp. 139–183, Paris (1990)
Petitot, Jean: Les modèles morphodynamiques en perception visuelle. *Visio*, 1, 1, spring: Représentation et cognition, pp. 65–73, Canada (1996)
Rader, Melvin: Empathy and Abstraction. Introductory Note. In: Rader M. (ed.) A Modern Book of Esthetics. An Anthology, 3rd edition. Holt, Rinehart and Winston, New York, Chicago, San Francisco, Toronto, London, pp. 367–370 (1960)
Rader, Melvin: The Experiences of the Beholder. In: Rader, M. (ed.) A Modern Book of Esthetics. An Anthology, 5th edition. Holt, Rinehart and Winston, New York, San Francisco, Dallas, Montréal, Toronto, London, Sidney, pp. 331–335 (1979)
Saint-Martin, Fernande: L'inscription du sujet thymique dans l'énonciation visuelle. Discours social/Social Discourse, II, 1–2, spring-summer, pp. 121–130 (1989)
Vischer, Robert: Über das optische Formgefühl: Ein Beitrag zur Aesthetik. Hermann Credner, Leipzig (1873) [English trans.: On the Optical Sense of Form: a Contribution to Aesthetics. In: Mallgrave, H.F.; Ikonomou, E. (eds.) Empathy, Form, and Space. Problems in German Aesthetics, 1873–1893. The Getty Center for History of and the Humanities, London (1993). New edition: Santa Monica, pp. 89–123 (1994)]
Wölfflin, Heinrich: Prolegomena zu einer Psychologie der Architektur, Inaugural-Dissertation der hohen philosophischen Fakultät der Universität München zur Erlangung der höchsten akademischen Würden. Kgl. Hof- & Universitäts-Buchdruckerei, Munich (1886). [English trans.: Prolegomena to a Psychology of Architecture. In: Mallgrave, H.F.; Ikonomou, E. (eds.) Empathy, Form, and Space. Problems in German Aesthetics, 1873–1893. The Getty Center for History of and the Humanities, London (1993). New edition: Santa Monica, pp. 148–190 (1994)]
Worringer, Wilhelm: Abstraktion und Einfühlung (1908) [English trans.: Abstraction and Empathy. A Contribution to the Psychology of Style. International Universities Press, New York (1953)]

Chapter 5
Interlocking Spaces in Art

Modern and contemporary art has often been seen as the cradle of new aesthetic perceptions and conceptions of space. Developed, according to the declarations of the avant-gardes, almost in opposition to production from the Renaissance onwards, modern art abandoned the Albertian model of representation and was acknowledged, by that, for the overcoming of an unchanging and homogenous spatial organisation. In this perspective, it might even be recognised as having the aesthetic primacy of the spatial complexity of art. Yet, in the following pages I would like to point out that it is appropriate to nuance these few statements, and the supposed break with previous achievements. Although I have also stressed that modern art implies a new awareness of some aspects of creation, in particular, the emergence of a psychophysics of perception in the late nineteenth century, I would now like to highlight the existence of an overlooked continuity with early modern[1] spatiality, beyond the changes that have occurred.

With this question in mind, I will then outline a short journey through spatial theories in art and gather some disparate avenues that have been formulated and could contribute to a semiotic analysis of the works of art's complexity. This is a research journey that has just begun, not yet delimited, for which, without any mannered rhetoric, I kindly ask the reader to be tolerant with the current state of research. This paper[2] does not intend to present consolidated findings, but seeks to stimulate discussion about art spatiality. Works of art, especially through the images

[1]Early modern refers here to the unevenly defined period from the Renaissance to the nineteenth century.

[2]A first version of this paper has been presented in the Séminaire Intersémiotique at the Maison de la Recherche, Paris, on 4 March 2009. I especially thank the session moderator Jean-François Bordron and the distinguished public of scholars attending the presentation for their valuable remarks, in particular, in the disorder of my memory, Claude Zilberberg, Anne Hénault, Bernard Bouteille, Ivã Lopes and Luisa Ruiz Moreno. The outcoming French article has been published with the title "Réflexions sur la complexité spatiale de l'art" in the academic journal *Nouveaux actes sémiotiques* (Caliandro 2009).

S. Caliandro, *Morphodynamics in Aesthetics*, Lecture Notes in Morphogenesis,
https://doi.org/10.1007/978-3-030-29631-5_5

conveyed by them,[3] most often articulate various spaces. Not only do they occupy space or, better still, participate in defining it, but above all they present and represent different spaces *in*, *through* and *from* them. Thus, one and only image can activate several spaces and become the nodal place, for example, of physic, perceptual and/or symbolic spaces, even inconsistent with one another. From my point of view, speaking of images as for works of art does not mean flattening the latter to a purely visual dimension. This perspective does not pretend to elude the question of the various sensorialities solicited by works of art, nor to diminish their material thickness. To consider the image in these terms is to bring to the fore the valuable aspect of art as *medium*, to see its potential to intertwine even conflicting spaces.

It seems crucial to start this study by going back to the foundations of early modern representation, which was supposed to have developed in the Renaissance period a rigorous technique of space construction. The linear perspective, produced by the adoption of a single and fixed point of view and by the tracing of a system of converging vanishing lines, would have generated in the picture, a space conceived as homogenous and measurable. This interpretation, very widespread in art history especially after Erwin Panofsky's seminal work, partly overlaps with the attitude of many modern aesthetics for which early modern representation was the model to reject in order to found new conceptions of space. The forsaking of the Renaissance perspective system would have even disrupted the artistic spatiality, henceforth freed, according to Fernande Saint-Martin, from "an abstract geometric space, which does not correspond to the sensory experience, nor to the knowledge that physical sciences have now reached on the nature of reality" (Saint-Martin 1980, new edition: p. 131, own translation, and chap. VI on "La perspective euclidienne"). However, by better analysing the spaces at stake in early modern times, one must nevertheless discard the idea of one coherent and rigorous space that would have been determined by the use of the linear perspective. It thereby becomes possible to appreciate the spatial complexity of such artistic production. And, in spite of the ruptures that the avant-garde and even postmodern aesthetics have wanted to declare to stand out from it, some continuity can be found in this respect between early modern art and modern and contemporary creation.

Historical and Semiotic Studies on the Typology of Perspectives

Sensory apprehension of space and its representation and perception through an image are two distinct moments, which art theory should help to avoid overlaying too hastily. While optics and vision sciences have always provided reference points for the construction of a space in painting or, in general, in the arts that essentially involve

[3]By images, I mean, here as in the rest of the paper, both the perceptions and the reproductions (whether or not two-dimensional) of works of art. Clearly this meaning puts forward, in respect to works of art, their moment of perception and reception.

two dimensions (drawing, engraving, photography, etc.), it has long been known there are many ways of representing the space perceived, in particular, to depict the depth of the third dimension. In *Perspective as Symbolic Form* (1927), Erwin Panofsky studied this topic from a historical point of view, considering various systems of representation and construction of space prior to the conception of linear perspective. Pursuing this work from a semiotic point of view, Fernande Saint-Martin has brought together a wider range of ways to depict depth in the picture and has thus sketched a synthesis of more than twenty-five "systems of perspective", from an extensive variety of geographical and historical backgrounds (Saint-Martin 1987, pp. 141–184 esp. 164–182; English transl.: pp. 109–144, esp. 130–144). By moving from ancient to contemporary art, from so-called decorative arts to children's production, from oriental art to aboriginal creation, she clearly relativizes the solution developed during the Renaissance and, in her enumeration, linear perspective and atmospheric perspective are only two cases among others.

At the same level as the perspectives already recognised in art history, such as the *axial perspective*—also known as *perspective with a vanishing axis* or *fishbone perspective*—and the *inverted perspective*, and parallel to the *cavalier perspective* and the *isometric* or *axonometric perspective*, both widely used in architecture and geometric drawing and whose use in art she also attests, Saint-Martin draws inspiration from plastic intuitions and the artists' writings in identifying other space techniques. For instance, she defines as *optical perspective,* the depth generated by the juxtaposition of various colours, tones, vectorialities, sizes, brightness and/or textures, which cause a perceptual floating of the planes (towards the front or in the background) with respect to what she calls, after Kandinsky, "the Basic Plane" of the picture (ibid., p. 168; English transl.: p. 145). Other forms refer more directly to artistic styles or movements, such as the *baroque perspective* and the *analytical and synthetic cubist perspective*, to the point that the particularly innovative work of an artist or a group seems sufficient to define a new type, such as the *checkerboard perspective*, having "a privileged role in the work of Mondrian" (ibid. p. 173; English transl.: p. 136), the *perspective of the Unism*, developed by W. Strzeminski to produce a homogenous mass of forms (cf. ibid. p. 171), or the *tachist perspective*, corresponding to "the application of the optical perspective to ensembles of informal regions" (ibid. p. 172; English transl.: p. 136). Without claiming to be exhaustive, the typology goes on very rich and mentions: the *focal perspective*, centred on the represented object which thus occupies almost the entire field; the *spheric perspective* developed on a curved space; the *parallel perspective*, proceeding by the juxtaposition of objects despite their actual distances; the *frontal perspective*, derived from the parallel one but by projection in vertical planes; the *anamorphic perspective*; etc. In this list, however, some types appear to be based solely on the point of view adopted, such as the *bird's eye perspective*, the *oblique perspective*, and the *perspective in high* as from very high mountains or from clouds (cf. ibid. p. 181). Moreover, some systems of perspective result from the composition or gathering of others, such as the *projective perspective*, in which "an ensemble of elements coexist, produced by heterogeneous points of views or distances" (ibid. p. 177; English transl.: p. 140), or the *folding back perspective*, in which a frontal, side or flat view is folded down on the ground (cf. ibid. p. 175). The question then arises as to the extent to which this typology constitutes a real classification, since some types are derived from others (e.g. the frontal, parallel and tiered—or *en étagement*—perspectives) or partially overlap with one another (such as the bird's eye perspective and the perspective in high, the folding back perspective and the projective perspective). For her part, the author will be ready to reply that the possible coexistence of several systems in one and only work of art raises no problem, especially since for her, according to Freud, contradiction does not exist and almost everything can coexist with everything.[4] As surprising as this position may seem, as the conclusions of this paper would also like to show, this is a provocative but still effective way of bringing to an end a logical, yet often simplifying line of thought about art theorisations.

Although Fernande Saint-Martin's typology may elicit some perplexities because of this heterogeneous coalescence of spatial techniques, it is nevertheless necessary to recognise her very high sensitivity in analysing topological relationships and their effects in the perceptual field. Furthermore, the inclusion of composite and hybrid perspective systems in the list is considerable, albeit they would paradoxically seem to confuse and weaken, at first sight, the value of her work. In particular, I would finally like to refer to the definition of what she calls the *arabesque perspective* (or "*à entrelacs*", that is, intertwined), in which the optical depth of colours and textures interacts with linear, parallel or crossed undulations that alternately dig and raise ("push and pull") the topological mass of the Basic Plane from front to back (cf. ibid. p. 170). As the author states, this technique, with its possible endless developments, ramifications and diversifications, has been widely used in Oriental, Persian and Egyptian and medieval art. The main reference, however, is Pollock's all-over dripping painting in which, according to Saint-Martin, the arabesque perspective would have come back without being recognised (cf. ibid.). Here as elsewhere in her book, there is a certain acrimony towards the *narrative* art of the Renaissance which, in order to impose its hegemony, would have rejected this and other perspectives, despite its own inability to render several "forms of sensorial experiences" (ibid. p. 179; English transl.: p. 141).[5] Marking a—somewhat preconceived—break between Renaissance's and modernist solutions, the author both reduces the importance of the linear perspective and values the richness and complexity of modern and contemporary art by ascribing to the latter two the heritage of different or past

[4]This is roughly what Fernande Saint-Martin argued in the doctoral course she held at UQAM in 1997–1998, following my request for clarification in presenting these systems of perspectives. At that time she had also introduced some minor changes to the typology published in *Sémiologie du langage visuel* in 1994.

[5]The quote is from the chapter concerning linear perspective, which, according to the author, is sometimes referred to as "legitimate, artificial, convergent, [or] central" (Saint-Martin 1987, p. 178; English transl.: p. 141). Giving it the label of "central" construction may explain the need to differentiate linear perspective from oblique perspective, although the latter is also based on *convergence* of lines and belongs, as the former, to the group of "The far distant perspectives" (ibid.). Finally, in classifying techniques able to represent objects in distance, Saint-Martin underlines that the "organizational structures [of linear perspective] prevent the possibilities of representing not only proxemic visual experiences but also other forms of sensorial experiences that unfold strictly in the proxemic distances", i.e. in close distance to the organism (ibid. p. 179; English transl.: p. 141).

civilisations or of creations long deemed as minor. Apart from these few debatable points, her contribution on composite and hybrid perspectives is thought-provoking. The idea of the possible co-presence of several perspective techniques deserves to be explored, especially since it could help to understand how different spaces, or a space with *n* dimensions, can articulate in one image.

Saint-Martin's argumentative position is in some respects similar to that of Erwin Panofsky when the latter explains that, in comparison with earlier systems of representation, the Renaissance managed an "extensive abstraction from the psychophysiological structure of space" "to construct an unambiguous and consistent spatial structure of (with the limits of the «line of sight») infinite extension" (Panofsky 1927, English transl.: p. 63). It is the structure of an "infinite, unchanging and homogeneous" space (ibid., English transl.: pp. 28–29) that Panofsky and Fernande Saint-Martin acknowledge in the linear perspective. Both underline the homogeneity of this spatial system, the former to describe it as a symbolic form pertaining to geometric construction and Euclidean mathematics, the latter to oppose it to the heterogeneous and articulated conception of a hybrid space in modern and contemporary art. Besides, Panofsky borrows this idea of a homogeneous geometric space from Cassirer, whom he quotes at length in his book:

> Hence homogeneous space is never given space, but space produced by construction; and indeed the geometrical concept of homogeneity can be expressed by the postulate that from every point in space it must be possible to draw similar figures in all directions and magnitudes. Nowhere in the space of immediate perception can this postulate be fulfilled. Here there is no strict homogeneity of position and direction; each place has its own mode and its own value. Visual space and tactical space [*Tastraum*] are both anisotropic and unhomogeneous in contrast to the metric space of Euclidian geometry.[6]

Panofsky thence posits the equivalence between Euclidean geometric space and the perspective space of the Renaissance.[7] A homogeneous and almost measurable construction would be at work in the production of that period.

This duality between a rationally constructed metric space and an inhomogeneous psychophysical space comes back in the writings of Jean-François Lyotard, who attacks "the Gestaltist organization of visual perception" as "the outcome of secondary rationalization" (Lyotard 1985, p. 156; English transl.: 153). Building on Barre and Flacon's essay *La perspective curviligne*, demonstrating that no regular shape, except the circle, can really be seen as is, and that education and habit lead the brain to rectify distortions in perception, Lyotard asserts: "Learning how to see is unlearning how to recognize" (ibid., p. 157; English transl.: p. 153).[8] And he urges to "give way to the pre-eminently figural space, to the field of vision which focalized attention represses, and which presents around the tiny area of clear vision (the

[6]The entire excerpt is from Ernst Cassirer, *Philosophie der symbolischen Formen* (1925), quoted in Panofsky (1927, English transl.: p. 30).

[7]Panofsky would not, however, have been the first to propose this equivalence, which can be found in (priest) Pavel Alexandrovich Florensky's essay on *reverse perspective* (Florensky 1920).

[8]Lyotard mentions in his note a long extract from: A. Barre and A. Flocon, *La perspective curviligne*. Flammarion, Paris. 1968.

foveal zone) a vast peripheral fringe of curved space" (Lyotard 1985, p. 157; English transl.: p. 154). From "*the irreversible heterogeneity between focal area and periphery*", the author points out, more than the significance of the margin, the emergence of the *difference*, "*dissymmetrical spatiality*" which incorporates disequilibrium into a structural system (ibid., p. 165; English transl.: pp. 161–162).

Panofsky is well aware of this problem when he states that the Renaissance perspective construction neglects or ignores these differences, not only that human vision is not based on one still eye, but also that the size and shape of objects in retinal image are affected by apparent changes. Thus, the lateral deformations, discussed by Renaissance theorists, were rectified in the geometric projection on a flat surface (cf. Panofsky 1927, English transl.: pp. 30–33). According to Panofsky, artists would have been led to overcome the discrepancy between *perspectiva naturalis* and *perspectiva artificialis* by correcting these incongruities (cf. ibid., English transl.: pp. 35–36).

Discrepancies between Art History, Art Theories and Works of Art

The *ideality* of the Renaissance perspective system is not anyway reflected in the artistic production of that period. While noting the consistent momentum for the enhancement of art science, Martin Kemp does not fail to recognise the substantial gaps between theoretical geometry and the systems used by artists to efficaciously reproduce human perception. From the first studies at the dawn of early modern time to works of art that advertently applied the new way of representing depth on the surface, not to mention the creations that deliberately thwarted this system, the art historian never ceases to recall the instinctive corrections and empirical adjustments that artists adopted to appropriate the technique and achieve the desired effects (cf. Kemp 1990, Italian transl.: p. 29).

Artists were indeed trying to avoid the impasse of the lateral deformation: in their artistic works and theoretical advices, both Brunelleschi and Piero della Francesca were careful not to exceed 53° the eye-view angle, or in any case be less than 90°, at least horizontally; similarly, the author of the Codex Huygens recommended not to use too close points of view, in order to avoid vertical accentuated foreshortenings, especially in the upper and lower extremities of the body.[9] But while these considerations were still related to the construction of linear perspective, Leonardo da Vinci began to put it into question by observing that the angles of vision for objects equidistant from one another gradually decrease in areas far from the centre, following his demonstration of the curved surface, and they cause monstrous deformations—to be avoided—for a beholder who would not position precisely in the place designated for the point of view. Moreover, from 1507–1508 with his treatise on the eye, Leonardo

[9]The author of the Codex Huygens, a mid-sixteenth century treatise in Northern Italy, seems to have been identified with Carlo Urbino from Crema. Cf. Kemp (1990, Italian transl.: p. 87).

will demonstrate that eye vision cannot be reduced to one point and that several pyramids of vision reach the organ; consequently, no edge of an object can be seen clearly, hence the appropriateness of his *sfumato* in contours of forms (cf. Kemp 1990, Italian transl.: pp. 55–64).

Whereas, in general, in their treatises, these artists were studying the most scientific and rigorous way to apply the new perspective technique, their artistic realisations deny a faithful and accurate use. This is evident not only in early works in which this spatial construction was still being developed—e.g. in *The Feast of Herod* (1423) Donatello set two distinct vanishing points, one for the floor and the scene in the foreground, the other for the architecture and the upper part of the bas-relief—but also in most subsequent perspectival representations. Even Paolo Uccello, one of the most fervent painters of foreshortened figures, seems to have drawn two vanishing points in the lunette of *The Nativity* (circa 1445), whose very poor state of conservation unfortunately prevents an adequate apprehension. He had also used this expedient in *The Deluge* (circa 1445) in which the double perspective construction with two vanishing points differentiates the temporality of the two scenes (during and after the deluge) in one and the same representation (cf. ibid., Italian transl.: pp. 47–49). Artists did not hesitate to bring together various spatial depth systems and amalgamate them to produce an efficacious perceptual impression. *The Assumption of Saint John the Evangelist*, sculpted in bas-relief by Donatello, or Mantegna's fresco in the ceiling of the Camera degli Sposi are further illustrations: the geometric tool hybridised with other disparate solutions wherein various perspectives—frontal, off-centred, in more or less accentuated low angle, etc.—share the space, therefore plural, of the picture.

Works of art of the Renaissance and, in general, in the early modern period then practise a spatial architecture already intertwined or, at least, an articulation such that the space defined by Panofsky as unchanging and homogeneous is rarely found there. Though from a theoretical viewpoint this kind of space was studied, and even designed, it was practically almost never carried out. It would be wrong to believe that such an absence of measurable and isotropic space results from an economy of practical means or from a lack of precision in the application of the model. First of all, because aesthetic intuition always prevailed over the mechanical execution of the geometric system. Second, because there are some treatises advocating the hybridisation of the model. Leonardo, as mentioned above, questioned linear perspective and juxtaposed this solution with the idea of a spatial view gradually changing as it approaches the limits of the visual field or, still, an atmospheric perspective fluidifying lines and contours and modifying colour tones to signify distance. Equally interesting, however, are Albrech Dürer's theoretical sketches[10] in which he geometrises the human body in volumes, projects it into space according different perspective orientations and deforms these blocks in an astonishing curvilinear undulation. The rectilinear rigour of the linear perspective was thus altered at the very moment this technique

[10]As Martin Kemp explains, these drawings, collected in *Vier Bücher von menschlicher Proportion* (Four books on human proportion), completed circa 1523 and published posthumously in 1528, concern the stereometry of the human body with regard to its movements in space.

began to spread. The adoption of several points of view and, consequently, several perspective systems in one and the same space—that of the painting or the perceptual field, already considered as heterogeneous—did not pose any problems for its aesthetic apprehension. Lodovico Cigoli's cupola (1510–1512)[11] in the Pauline chapel in Santa Maria Maggiore in Rome shows foreshortened figures projected according to a beholder located below the fresco in the centre of the chapel, while the figures on the border of the scene bear an apparent distortion and are represented almost in frontal view. Yet, most notably, the image of the Madonna does not conform to the foreshortened vision of the whole and rather takes on her full height when seen from a point of view located exactly at the entrance of the chapel. Clearly, the painting on architectural surfaces, where the visitor's gaze follows a travelling itinerary in accordance with the spatial and effective accessibility of the site, increased the possibilities of perspectival interweaving, which the baroque ceilings and the use of the *squadratura* would refine and make more and more articulated.

Going back to the gestaltist organisation of perception that Lyotard denounced, apparently it was mainly criticism and art theory that wanted to identify a *good form*, a kind of ideal order based on linear perspective, in the complexity of spaces elaborated in early modern art. If a secondary rationalisation took place, it is more a matter of art history than of works of art or, even, of artists' writings of that time. In addition, in his *Theory of /cloud/* Hubert Damisch has illustrated how painters succeeded in integrating incommensurable spaces into a work of art and how, through the use of the cloud—that is, the least easily measurable and most confusing element in any perspective system— artists introduced the transcendence of the figurable in figurative representation (cf. Damisch 1972). Furthermore, the same author has admitted the idea of *fold* in pictorial space, when analysing Caravaggio's *Narcissus* in comparison to Poussin's painting on the same subject (cf. Damisch 1996). In this sense, spatial complexity is not a prerogative of modern era and contemporaneity and their art; it is not only found in ancient and medieval art because of inexperience in designing an unchanging and homogenous space; but it crosses the productions of all periods. To depict spatiality is an old and current challenge, and only a simplifying geometric abstraction could reduce its articulation to a merely linear perception.

Research on the Complex Spatiality of Art

With regard to the positioning of the point of view, whether fixed or likely to move along a path pre-figured by the work of art, its determination undoubtedly contributes to further enrich the spatial complexity of the picture, since it relates presented and/or represented spaces to the beholder's space. In *La macchina della pittura*, Omar Calabrese has increased the double spatiality that Félix Thürlemann observed in painting, viz., the "simulated space" in the representation and the "planar topology" or flat

[11] Lodovico Cardi, also know as Cigoli, *The Virgin of the Immaculate Conception*, 1510–1512, Pauline or Borghese chapel in the church of Santa Maria Maggiore, Rome.

surface of the painting, with a third type of space: the space in front of the representation, where the beholder is positioned (Calabrese 1985, pp. 268–269, own translation, referring to Thürlemann 1981). This triple spatiality has thus made it possible, for Calabrese, to establish a triple correspondence, respectively: between semantics and simulated space (i.e. virtual space, the space of depth, the space beyond the canvas), between syntax and planar topology (surface space) and between pragmatics and space of reception. This three-pole articulation, with clear reference to the three semiotic functions, has been revised a few years later by the same author, who has proposed to differentiate rather four types of spaces in painting (cf. Calabrese 1987, Italian transl.: pp. 161–164, esp. 162). To the three above-mentioned spaces—1. The perspectival depth, *on the other side* of the canvas, 2. The depth *on this side*, in front of the canvas, 3. The surface of the picture as a geometric area that he continues to call topology—Calabrese adds: 4. The material surface or thickness of the painting. Of course, as he clarifies straightaway, this typology is only theoretical, since these spaces can be separated only for the purposes of the analysis. The appropriateness of distinguishing the fourth point is nevertheless debatable, since it could be embodied in the topological dimension of the work of art, at least if, instead of considering topology as purely planar, as Thürlemann also did, the term is used in accordance with its physical and mathematical meaning. Moreover, the distinction between the surface of the painting and its material surface or thickness does not lead at all to differentiate abstract art from figurative art or early modern representation, as Calabrese's article intended. In my opinion, this distinction stems from a very subtle and less schematic study developed by Louis Marin, from which Calabrese may have drawn inspiration. Anyhow, taking up again the pursuing of Marin's thoughts will help to understand how even the spatial articulation proposed by Calabrese cannot be taken in too rigidly, as the academic clarity of the article might suggest, and that on the contrary, he tries to untangle the strands of a very complex skein.

In "L'espace Pollock", analysing the production of this American painter, Louis Marin preliminarily distinguishes, on the one hand, the *space of the beholder* who looks at the painting from a fixed or mobile point of view and, on the other hand, the *space of the canvas* (in French: *espace du tableau*), understood as the support or medium which conveys what is displayed, i.e. the *plane* (cf. Marin 1980). To these two, he immediately juxtaposes the space of the painting or *space in the canvas* (*espace dans le tableau*), that is, the space constructed, for example, through representation: it is the illusory depth that representation digs into the surface of the canvas through the perspective. These three spaces are conceived dialectically or, better still, by a *dynamic* dialectic between the position of the beholder and his or her path, by a *spatial* dialectic between canvas and plane, and by a *material* dialectic between surface and depth. It is worthwhile to enhance this multi-dynamic articulation, since it maintains at all times a strict interrelation between these three spatial dimensions. But it is also important to be mindful to the floating between canvas and plane, from which I think Calabrese's distinction between the topological surface of the canvas and its material thickness has been drawn. For Marin, the canvas, in general, is not visible, occulted as is by the background, the furthest layer of the represented space.

At the same time, the plane is treated as it was transparent, as a veil interposed on the window of the representation. Thus, even a nonfigurative painting, such as that of Jackson Pollock, perceptually floats between canvas and plane; it cannot resume the transparent plane of the Albertian window, but is also not merely a painted support: it is the translucent thickness of an interstitial space in which, as Marin explains, the eye glimpses; it is a world of spatiotemporal intervals teeming with countless differences. Instead of trying to distinguish, even theoretically, these intertwining concomitant spaces, I believe that the strength of Marin's analysis lies exactly in having conceived them in their complexity. Similarly to the intertwined perspective proposed by Saint-Martin, it is not the sum of the single perspectival ramifications or developments that defines the spatiality of the whole, but the actual fact that these spaces are intrinsically interlocking.

On the Possible Coexistence of Inconsistent Spaces

To conclude, at least temporarily, this journey between visual theories and semiotics of art, I would like to return to Lyotard, but passing briefly through Klee. In a previous study on this artist's production, I pointed out that, for Paul Klee's plastic thinking, one of the ways to render "the living form" (Klee 1956, English transl.: p. 60) was to "force the third dimension into the flat plane" (Klee 1957, English trans.: p. 124), viz. not to represent the third dimension as such, but to depict the coexistence, even conflicting, between it and the two-dimensionality of the picture.[12] In René Thom's words, it means the possibility of sending a space into another space of lower dimension. The mathematician has illustrated it with the sleeve of his jacket: when he compresses it, the fabric makes folds appear; that is to say, the fabric accepts this spatial constraint except in some points which manifest the singularity of their own dimension. The difficulty of thinking of several types of spaces simultaneously or, even, of spaces of various dimensions in one single space, can be simplified by the example given by Margaret Wertheim in her widely disseminating book on the history of conceptions of space, which I will freely cite here. For her, following the ideas of the mid-nineteenth century mathematician Bernhard Riemann, a world conceived as a sheet of paper, where two-dimensional inhabitants live, would not cease to be such if the paper were folded and massed in a ball. For those beings, knowing only the second dimension, their world would continue to appear planar, yet it would not longer behave as such (cf. Wertheim 1999, Portuguese transl.: p. 150).

Without having recourse to the support of non-Euclidian sciences, Jean-François Lyotard argues that the censorship of the curvature of space, carried out by a rationality that imposes the *good* point of view, can be grasped through Freud's metaphor of the unconscious space.[13] In *Civilization and Its Discontents*, "by a flight of imagina-

[12]Cf. Caliandro (2001–2002, p. 31), translated herein as Chap. 2.

[13]Lyotard refers again to the "peripheral curvature" of the eye vision, analysed *supra*. (Lyotard 1985, p. 182; English transl.: 179).

tion" Freud invites to "suppose that Rome is not a human habitation but a psychical entity with a similarly long and copious past—an entity, that is to say, in which nothing that has once come into existence will have passed away and all the earlier phases of development continue to exist alongside the latest one." (Sigmund Freud quoted in Lyotard 1985, note n. 61 pp. 184–185; English transl.: note n. 64 p. 434)[14] In particular, he gives the example of the site of the Pantheon, on which it would then stand not only this monument as is today but also the original edifice erected by Agrippa, the later church of Santa Maria sopra Minerva, as well as the ancient temple on which it was built. "And the observer would perhaps only have to change (*Änderung*) the direction of his glance or his position in order to call up the one view or the other." If for Freud it makes no sense to pursue this vision, this "phantasy" which "leads to things that are unimaginable and even absurd", it is because in our perception of an historical sequence we are not able to imagine the spatial concomitance of different forms: "the same space cannot have two different contents" (Lyotard 1985, p. 184; English transl.: p. 435). For Freud as for Lyotard, this is indeed the specificity of the space of the unconscious.

Without necessarily continuing the psychoanalytical approach of the authors mentioned here (Freud, Lyotard, but also Saint-Martin and, in his style, Damisch), it is interesting to recall a sentence, currently out of date, written by Freud just before the quotation given above: "Where the Coliseum now stands we could at the same time admire Nero's vanished Golden House." (Freud quoted ibid.) In his flight of fancy, the psychoanalyst could not expect that this architecture was still there, buried under a hill facing the Colosseum (or Coliseum), and would have been accessible to the public in the recent past. Luckily, after spending years in this ancient quarter of Rome, I have visited this historic monument[15] and experienced the change that its rediscovery has brought about in the perception of the city. The Domus Aurea (or Golden House) has indeed retained all its monumental grandeur and literally plunges the visitor into a world whose living conditions are quite different from contemporary ones. Thus, like many Romans, I have felt the difficulty of these two types of irreconcilable and yet coexisting spaces: the hill in its current urban setting and the huge architecture hidden in its belly. Unlike the Pantheon, which in its present state, contrary to Freud's imagination, displays the cultural stratifications and transformations undergone over time, the perception of the Domus Aurea site imposes an unusual and, indeed, improbable perspective. Henceforth, two types of spaces are interdependent in this site; and perhaps, by an interlacing anisotropic vision or by

[14]Lyotard refers directly to the German text of Sigmund Freud, *Das Unbehagen in der Kultur* [1930], GW, vol. 14, pp. 427–428.

[15]Excavations were rather sporadic over the centuries and progressed mainly in the twentieth century. Then the monument was closed for a long time for works since the early 1980s and reopened some years ago. Visits, limited to small groups and on reservation, have been interrupted again for many years, due to security reasons, and partially authorised from 2015.

a discontinuous leap in the direction of the gaze, it would no longer be possible to think of one without the other.

The break that modernism and then postmodernity have claimed to mark in relation to the past does not quite concern the spatiality of art. The latter has allowed the evolution of its forms, its perspective systems and its various ways of interrelating with the subject. But a continuity exists between early modern art and modern and contemporary art with regard to the complexity of the spaces involved. Although art historians have described the Renaissance perspective space as unchanging and homogeneous or, better still, Euclidean, the analysis of the artistic production and theories of that period reveals, instead, a heterogeneous and articulated use of space. It is a methodological problem to have simplified this point in order to increase the intelligibility of some historical elements or phenomena. Yet, some art theorists and semioticians have outlined interesting possible avenues for considering complexity as an essential moment not only for the aesthetic manifestation and apprehension of the work of art but also for the understanding of its meaning. It is desirable that research succeeds in developing a theorisation capable of analysing concomitant spaces, even when irreconcilable, in order to avoid that the explanatory or descriptive requirements of the analysis unravel the various entanglements the spatiality of art produces through its images.

References

Calabrese, Omar: La macchina della pittura. Pratiche teoriche della rappresentazione figurativa fra Rinascimento e Barocco. Laterza, Rome-Bari (1985)

Calabrese, Omar: Problèmes d'"énonciation abstraite". Actes Sémiotiques. Bulletin, 44. Paris (1987) [Italian transl.: Problemi di "enunciazione astratta". In: Corrain, L., Valenti, M. (eds.) Leggere l'opera d'arte. Dal figurativo all'astratto. Esculapio, Bologna, pp. 161–164 (1991)]

Caliandro, Stefania: Le point gris de Paul Klee. Visio, 6, 4, winter, Quebec, pp. 21–41 (2001–2002)

Caliandro, Stefania: Réflexions sur la complexité spatiale de l'art. Nouveaux Actes Sémiotiques, "Recherches sémiotiques", Université de Limoges: PULIM, 14 December 2009. https://www.unilim.fr/actes-semiotiques/1721 (2009)

Damisch, Hubert: Théorie du nuage. Pour une histoire de la peinture. Éditions du Seuil, Paris (1972)

Damisch, Hubert: Narcisse Baroque? In: Buci-Glucksmann, C. (ed.) Puissance du Baroque. Les forces, les formes, les rationalités. Éditions Galilée, Paris, pp. 29–42 (1996)

Florensky, Pavel: Reverse Perspective. [1919] (1920). New edition in: Beyond Vision. Essay on the Perception of Art. Reaktion Books, London, pp. 197–272. http://uberty.org/wp-content/uploads/2015/12/pavel-florensky-beyond-vision-essays-on-the-perception-of-art-1.pdf (2002)

Kemp, Martin: The science of Art. Optical themes in western art from Brunelleschi to Seurat. Yale University Press, New Haven, London (1990) [Italian transl.: La scienza dell'arte. Prospettiva e percezione da Brunelleschi a Seurat. Giunti, Florence. (2005)]

Klee, Paul: Das bildnerische Denken, collected and edited by Jürg Spiller. Schwabe & Co, Verlag, Basel, Stuttgart (1956) [English transl.: Notebooks. Volume 1. The thinking eye. Lund Humphries, London. 1961]

Klee, Paul: Tagebücher von Paul Klee 1898–1918, edited by Felix Klee. Verlag M. DuMont Schauberg, Cologne (1957) [English transl. authorized by Felix Klee: The Diaries of Paul Klee, 1898–1918. Berkeley. University of California Press, Los Angeles, London (1964)]

Lyotard, Jean-François: Discours, figure. Klincksieck, Paris (1985) [English transl.: Discourse, Figure. University of Minnesota, Minneapolis (2011)]

Marin, Louis: L'espace Pollock. Cahiers du Musée National d'Art Moderne, 10, pp. 316–327 (1980) [Italian transl.: Lo spazio Pollock. In: Corrain, L. (ed.). Semiotiche della pittura. I classici. Le ricerche. Meltemi, Rome, pp. 207–224 (2004)]

Panofsky, Erwin: Die Perspektive als 'symbolische Form'. Vorträge der Bibliothek Warburg 1924–1925. Leipzig, Berlin, pp. 258–330 (1927) [English transl.: Perspective as Symbolic Form. Urzone, New York (1991)]

Saint-Martin, Fernande: Les fondements topologiques de la peinture. Essai sur les modes de représentation de l'espace, à l'origine de l'art enfantin et de l'art abstrait. Hurtubise HMH, Montreal (1980). New edition: 1989.

Saint-Martin, Fernande: Sémiologie du langage visuel. Presses de l'Université du Québec, Sainte-Foy (Quebec) (1987). New edition: 1994 [English transl.: Semiotics of Visual Language. Indiana University Press, Bloomington and Indianapolis. (1990)]

Thürlemann, Felix: La double spatialité en peinture: espace simulé et topologie planaire. Actes Sémiotiques. Bulletin, 15. Paris, pp. 34–46 (1981) [Italian transl.: La doppia spazialità in pittura: spazio simulato e topologia planare. In: Lucia Corrain, Mario l'opera d'arte. Dal figurativo all'astratto. Esculapio, Bologna, pp. 55–64 (1991)]

Wertheim, Margaret: The Pearly Gate of Cyberspace: a history of space from Dante to the Internet. W. W. Norton, New York, London (1999) [Portuguese transl.: Uma história do espaço de Dante à Internet. Jorge Zahar Editor, Rio de Janeiro (2001)]

Chapter 6
Moscow I by Wassili Kandinsky (or, The Psychophysics of Perception in Art)

The psychophysical approach, which first emerged in the mid-nineteenth century and contributed to many thoughts in aesthetics and theories of perception in the second half of the century, succeeded in influencing artistic research at the beginning of the following century, thus also preparing the ground for the contemporary era. In particular, contrary to the connotation of formalist that has often been assigned to it, an organic conception seems to support the emergence of abstract art. A fine relationship can likely be established between these theories and artistic experimentations, and mark the intertwining of empathy with abstraction, through a passage that entails some important retours, as it will be discussed here, towards figuration. A work of art by Wassily Kandinsky, considered by critics, broadly speaking, as the 'inventor' or at least the theorist and major promoter of the abstract movement, will essentially provide the scope for this study.[1]

It is advisable, however, to quickly point out that abstract art, according to its historical definition, cannot be restricted to the sole question of going beyond mimesis, an issue that finally permeates the entire history of art long before the contemporary era. The problem of iconicity—and the questioning of its foundations (by the semi-figurative), or its real reversal—should rather be framed within the new conception that this movement embodied, in its disparate tendencies. The representational value was indeed most often replaced by a renewed attention to the aesthetic relationship that the work of art establishes with the perceiver. In this respect, it is easy to understand the embarrassment of the main proponents of abstract art who preferred for their work to express, respectively, in terms of "concrete art" (Kandinsky), "neoplasticism" (Mondrian) or "new realism" (Malevich). An article written by Kandinsky in 1935 explicitly states:

---

[1] The original French version of this paper, presented at the Congress of the International Association for Visual Semiotics held in Quebec City in 2001, has been published in the academic journal *Visio* (Caliandro 2004).

S. Caliandro, *Morphodynamics in Aesthetics*, Lecture Notes in Morphogenesis,
https://doi.org/10.1007/978-3-030-29631-5_6

> The expression "abstract painting" is not appreciated. And it is right as it does not mean much, or at least it is confusing. This is why abstract painters and sculptors in Paris have tried to create a new expression: they say "*art nonfiguratif*", equivalent to the German expression "*gegenstandslose Kunst*".
>
> The negative elements of these terms ("*non*" and "*los*") are not clear: they exclude the "object" without putting anything in its place. For quite some time an attempt as been made (as I did before the war) to replace "abstract" by "absolute". Actually, it would not be much better. In my opinion, the best term would be "concrete" [*reale Kunst*], since this art juxtaposes a new world, of spiritual nature, to the outer world. A concrete world. But the old name of abstract art has already become established. (Kandinsky 1935, new edition: p. 172, own translation[2])

The debate on the psychic effects produced by shapes and colours led artists to assign a value of origin to the plastic mediums, as it were a new genesis. Cosmogenetic ideas appeared in the attitudes of the major painters and artists who founded abstraction (Klee, Malevich and, as I will try to show, also Kandinsky in a way). The research they developed into plastic means, instead of directing them towards a simple formalism, led them to go deeper into the effects, dynamics and "tensions" (to use a term dear to Kandinsky[3]) that arise in perception. They were thus aiming at spiritual values,[4] even morphogenetic ones for those who, such as Paul Klee, were concerned with rendering the "living form" (Klee 1956, English transl.: p. 60). Kandinsky, for his part, conceived the pictorial field in terms of "basic plane": the sense emerges in the painting through the "tensions" that are created "optically and psychologically" from the topology of the elements on the surface (Kandinsky 1975, pp. 72–75, own translation). That this is not simply a return to two-dimensionality is

---

[2]Here is the original excerpt in German:

> Der Ausdruck "Abstrakte Kunst" ist nicht beliebt. Und das mit Recht, da er wenig sagend ist, oder mindestens verwirrend wirkt. Deshalb versuchten die Pariser abstrakten Maler und Bildhauer einen neuen Ausdruck zu schaffen: sie sagen "art nonfiguratif".
>
> Gleichbedeutend mit dem deutschen Ausdruck "gegenstandslose Kunst". Die Negationsteile dieser Worte ("non" und "los") sind nicht geschickt: sie streichen den "Gegenstand" und stellen nichts an seine Stelle. Schon seit längerer Zeit versuchte man (was auch ich noch vor dem Krieg tat) das "abstrakt" durch "absolut" zu ersetzen. Eigentlich kaum besser. Der beste Name wäre meiner Meinung nach "reale Kunst", da diese Kunst neben die äußere Welt eine neue Kunstwelt stellt, geistiger Natur. Eine Welt, die ausschließlich durch Kunst entstehen kann. Eine reale Welt. Die alte Bezeichnung "abstrakte Kunst" hat sich aber bereits eingebürgert.

[3]Cf. for instance Kandinsky (1926, French transl.: pp. 41–42, 66, 101–102 and 135–136).

[4]Let's think of Malevich's suprematism, Mondrian's theosophical values, or Kandinsky's essay, one of his first, which is precisely entitled *Über das Geistige in der Kunst, insbesondere in der Malerei* (*Concerning the Spiritual in Art: Especially in Painting*, published at the end of 1911 and dated 1912).

clearly observable on the attention he pays, as other abstract artists do, to the depth effects produced by colour. That an organic conception of the field is at the basis of his research remains, however, a point that criticism does not seem to have sufficiently considered. A look at the paintings Kandinsky made from his move towards abstraction in 1910–1911,[5] however, readily shows how his theories completely permeate his pictorial practice. Moreover, such a psychophysical conception seems to go beyond the limits of a strictly nonfigurative production to feed more generally his work.

The series the artist devotes to Moscow supports these considerations. The painting *Moscow I (Red Square)* of 1916 will be the main reference, bearing in mind that it is part of a set of studies—apparently never completed into a final and decisive work—that Kandinsky undertook around that date. This work, therefore, belongs entirely to his Russian period, still little known by the general public, and whose production, which he had to leave behind when he fled the Soviet Union, have hardly been exhibited abroad.[6] This Russian period, which Will Grohmann has hastily and pejoratively described as an "intermezzo" (Derouet and Boissel (eds.) 1984, p. 145, own translation), extends from 1915, the year the war forced Kandinsky to leave Germany and return to his homeland, to 1921, when the Soviet New Economic Policy was adopted, imposing more and more constraints on artists (art was henceforth subject to propaganda, national ideology and production; the following year abstraction was officially banned).[7] Although the artist complained that he had painted relatively little during this period,[8] his drawings and canvases nonetheless attest to an interesting renewal in his painting. After the great impulse towards abstraction of the previous period, he seems to embrace a return to figuration or, at least, to the semi-figurative by a very free and conscious choice, consistent with his approach.

---

[5]Critics are not unanimous on the exact moment of this move: the 1910 watercolour that some supposed to be his first abstract work has been dated by Kandinsky much later; so, others prefer to indicate as reliable the following year, when the artist realises the abstract oil *Bild mit Kreis* (*Picture with a circle*, 1911).

[6]An exhibition in Rome—from 7 October 2000 to 4 February 2001—tried to fill this gap. Cf. Comerlati (ed.) (2000).

[7]A three-month stay in Sweden (December 1915 to March 1916) is also included in this period.

[8]Cf. Derouet in: Derouet and Boissel (eds.) (1984, p. 145).

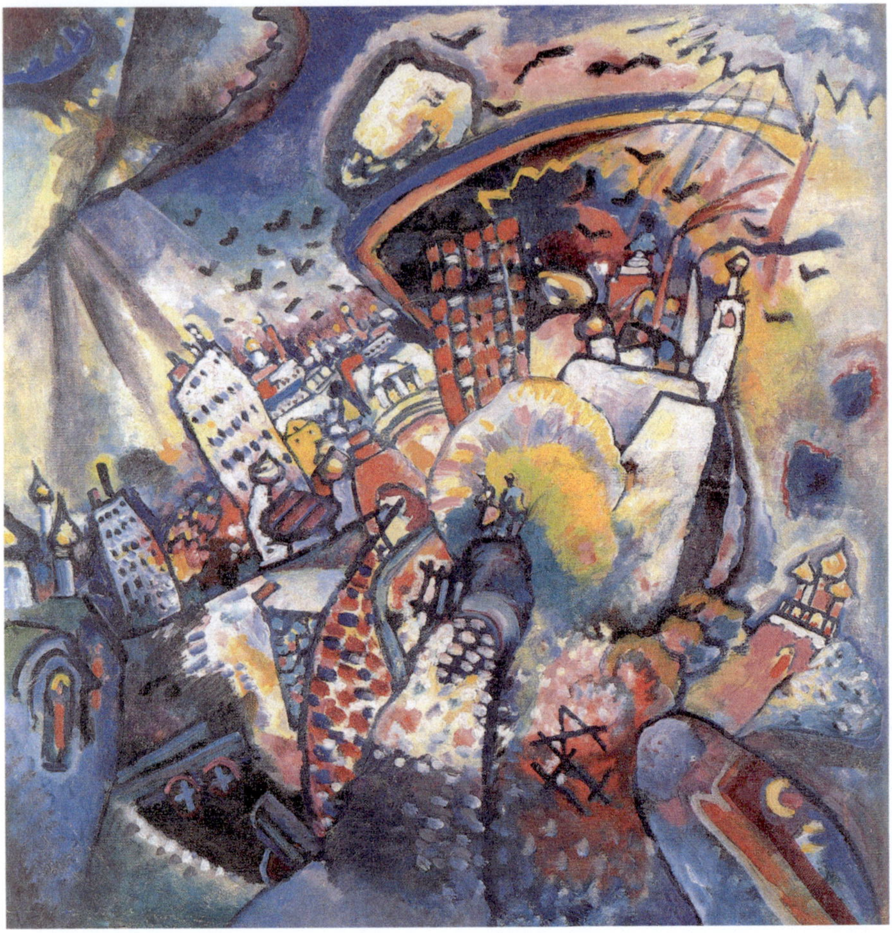

Wassily Kandinsky. *Moscow I (Red Square)*. 1916. Oil on canvas. 51.5 × 49.5 cm. The State Tretyakov Gallery, Moscow

This corpus of studies seems in general to provide a referential and contextual anchor for his painting. Its iconicity, however, gives new significance to his optico-psychological conception of art. In that respect, there is indeed no break from the theoretical point of view. In his lecture, "On a method for working with synthetic art" presented at the Russian Academy of Artistic Sciences in 1921, Kandinsky extends the line of thinking pursued since *Concerning the Spiritual in Art*:

> Inasmuch as works of art are, ultimately, intended to exert this or that effect on the individual, the ultimate aim of researching the properties of the elements in themselves is to establish this effect as universally as possible. It is this point of view that is used to study the physical properties of the elements, their physiological actions, etc.
>
> [...] Thus, for example, in the science of painting, the colour as main element is studied in the system of its entire set of properties: physical properties and its essence, psychological influences through the eye and through other means (coloured baths), psychic emotions in

> normal and pathological cases, capacities to induce associations in the domain of senses and representations, etc. (Kandinsky 1921, in ibid.: p. 158–159, own translation).

In 1921, Kandinsky indeed becomes the vice-president of the Russian Academy of Artistic Sciences and the head of the Physico-psychological Department.[9] Apparently, the idea of working on figuration again, which the artist had been practising for some time, does not contrast at all with its commitment to delving into the psychic effects conveyed by the physical elements of the painting. On the contrary, the introduction of iconic motifs seems to add new emotional resonance to the use of plastic means. It is also interesting to note that, especially shortly after the return to his hometown, he devotes several paintings to subjects that reflect not only the contextual, but also an emotional and very personal value. This is the case of *Moscow, view from the apartment window* (circa 1916) and *Zubovskaya Square* (circa 1916). For the latter, he had already sketched a study (*Untitled*) in Indian ink in 1915[10]; it really corresponds to the view from the building he had bought to ensure certain stability to his ménage and for which he had made a strong financial commitment.[11] These almost 'intimate' views that he paints of Moscow thus reveal a positive emotional charge that will soon clash with factual hardness when the artist has to cope with a very serious economic crisis.[12] But this affectivity also dominates in the visions he gives of Moscow, still in 1916 (see the two versions *Moscow I (Red Square)* and *Moscow II*, painted circa 1916). The relationship he felt with his city, the traditions of his country, its past, was always very intense. In *Reminiscences*,[13] he dwells at length on the images of his youth and his journeys in the rural world:

> I shall never forget the great wooden houses covered with carvings. In these wonder houses I experienced something that has not repeated itself since. They taught me to move within the *picture*, to live in the picture. [...] When finally I entered the room, I felt myself surrounded on all sides by the painting, into which I had thus penetrated. The same feeling was dormant in me, until then completely unconscious, when I was in the churches in Moscow,

---

[9] According to Nina Kandinsky's recollection, he was not allowed to hold the office of president of the Russian Academy of Artistic Sciences because he was not registered with the party.

[10] According to the catalogue of the Centre Pompidou, this study, inv. 336, dated on the back by Nina Kandinsky, bears in the lower left the inscription "Zubovskaya Square" in Russian; cf. Derouet and Boisse (eds.) (1984, p. 161). The artist painted several versions of this view (see *Moscow, Zubovskaya Square III*, circa 1916).

[11] Cf. ibid., pp. 148–149, for details on this placement as head of household and to appreciate the personal investment Kandinsky made in it beyond the emotional upheaval he experienced at that time (in March 1916 he definitively separated from Gabriele Münter and one year later, in February 1917, he met his new wife Nina; cf. ibid., pp. 146 and 150).

[12] Not only did Kandinsky have to abandon the project of building a villa for himself, but in 1917, in a campaign of expropriation following the October Revolution, the artist, no longer able to meet his mortgage obligations, sold his building and finally lost everything he had acquired. Kandinsky will mention several times the difficulties and hardships of these years in Moscow. Cf. ibid., p. 148.

[13] This book, originally published with the title *Rückblicke* in Munich and Berlin in 1913, was revised and refined in detail by the artist himself for the Russian edition in 1918. An early partial translation into English was published by the Solomon R. Guggenheim Foundation with the title "Retrospects by Wassily Kandinsky" in 1945, upon the occasion of the Kandinsky memorial exhibition in New York.

> and particularly in [the Kremlin Cathedral][14]. […] Of course, every time the impression was coloured entirely different, because completely different parts formed this impression. (Kandinsky 1913, English transl.: p. 27 modified)

This feeling awakens and renews itself when Kandinsky engages in pictorial work. His art explores how to restore the significance of the aesthetic, emotional and cognitive relationship that connects the picture to the perceiver. His painting, of which he highlights the *concrete* nature, cannot be grasped outside this relationship with the context of creation and reception. Not only does the work of art reveal the perception of the world that the artist felt, but it is now aimed at bringing the viewer into it, the latter being involved in it, in a similar way, through aesthesis. *Moscow I (Red Square)* leads the beholder's gaze to enter the painting. The oblong shapes, which from the lower part appear to converge towards the centre of the canvas, are like vectors inviting us to penetrate towards the centre of the picture. In particular, this shape that from the lower middle ends just below the yellow circular zone outlines an optical depth that encourages kinaesthetic projection. It is a kind of passerelle connecting the viewer with the two painted figures at its end. Placed in the non-geometric but rather perceptive centre of the canvas, these figures then assume an 'enunciative' value in relation to the pictorial work surrounding them; or at least they constitute the perceptual anchor point from which the subject's sensoriality explodes in all directions. From this yellow zone that the two personages contemplate, the perception seems to radiate towards several parts: the brightness of the yellow zone indeed induces to locate the other light sources in the upper right and left; its intense colour gives rise to the resonance of these almost primary colours which constitute nearly everywhere the raw material of this pictorial world. At the same time, the central figurative anchorage leads to search for the other iconic elements in the picture: the urban landscape, dotted with domes atop the buildings, the temple, the churches, the skyline of this citizen environment, crowned by the rainbow. And also, the black marks that fly over the city, the black hole with the two small crosses at the bottom left, the red spots lining the left path: plastic elements that become figurative in the context; one can guess birds, perhaps a tomb, human beings… This journey towards the iconic does not cease to reverse, though. Perception follows various dynamics, it scatters, jumps between different planes and depths of the picture, returns to the materiality of the pictorial means and their sensory evocations. It proceeds discontinuously, from the centre to the periphery but with unpredictable bounces between parts even far away in the picture. As Kandinsky writes when reminiscing about traditional Russian houses, highly decorated and full of icons, the subject learns to "move within the *picture*, to live in the picture" (ibid., cf. *supra*):

> I did a lot of sketching of these tables and various ornaments. They were never paltry and were painted with such strength that the object dissolved within them. This impression, too, only became clear to me much later.

---

[14]In the Russian edition Kandinsky substitutes with: "the Cathedral of the Assumption and the Church of Vasily's the Blessed".

> It was probably through these impressions, rather than in any other way, that my further wishes formed themselves within me, and thus the objectives of my own art. I have for many years sought the means of letting the viewer ["stroll"][15] within the picture, forcing him to become absorbed in the picture, forgetful of himself. (Ibid., English transl.: p. 28 modified)

The change that the artist introduces in the Russian edition (1918) of this essay—"move" instead of "stroll" within the picture—underlines the emotional charge that these perceptual dynamics entail. From the outset, the subject participates in the very genesis of the pictorial world. Similarly to these meaningful optico-psychological effects created by tensions and plastic elements, figurative objects seem to refer to an affective symbolism rooted in the artist's original culture, in his life experience. It then comes back to mind the above-mentioned passage on the Kremlin Cathedral that *Moscow I (Red Square)* cannot avoid recalling. However, it is the constellation of domes that symbolically evokes here the impression of moving in a painting as Kandinsky experienced it when he entered certain churches.

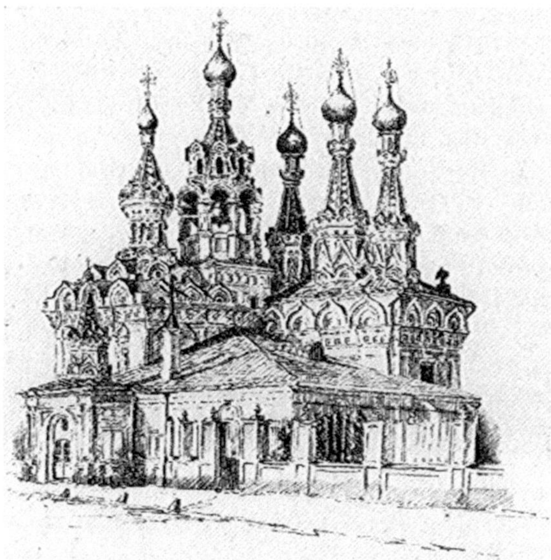

Wassily Kandinsky. *Untitled (Church of the Nativity of the Virgin)*. [1886]. Indian ink on paper. 19.9 × 15.1 cm. Centre Pompidou, Paris

[15] In the Russian edition: "move".

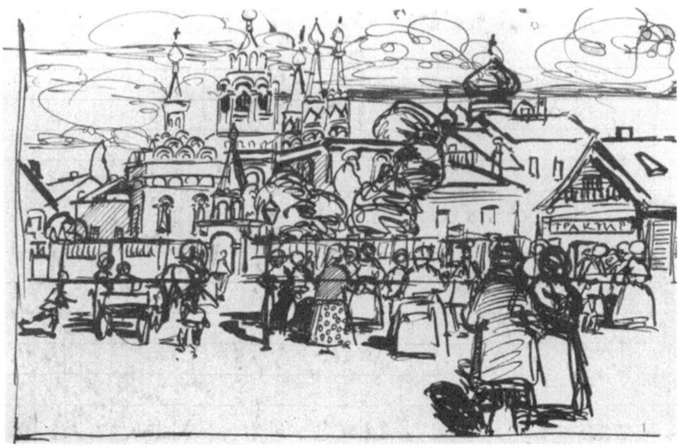

Wassily Kandinsky. *Untitled*. n.d. [inv. 345]. Indian ink on paper. Centre Pompidou, Paris

These churches may be also identified with some of the churches that the artist had repeatedly depicted since the beginning of his career (see *(Untitled. Church of the Nativity of the Virgin)* dated, by Nina Kandinsky, of 1886). In addition, very early on in his production, Kandinsky introduced small figures, namely a couple, that facilitate the beholder's empathic relationship with the represented scene (see the Indian ink *(Untitled)*[16] and *Moonrise*, 1904). Moreover, in the series of studies concerning *Moscow I* the position of the two personages is attentively studied: the posture apparently from the front that they assume in *Improvisation Gorge* (1914, see also *Sketch for Improvisation Gorge*, 1914) is first shifted towards the centre of the work of art (see two other studies *(Untitled)*[17]), then turned around, the back facing the beholder who thus find his or her own position represented within the painting *Moscow I*. Compared to their first achievements (see *Moonrise* and *Improvisation Gorge*), these small figures are also very schematically sketched and consequently acquire a generalising, de-individualised appearance. The beholder's sensitivity is solicited, the identification process is facilitated or, at least, the empathic immersion in the picture is encouraged. Hence, the subject, as well as the personage, enjoys through aesthesis the pictorial universe in which it participates. As the artist wrote, the viewer is invited, even forced, to enter the painting, to "move", to "become absorbed" in it, "forgetful of himself" (ibid., cf. *supra*).

[16]The catalogue of the Centre Pompidou compares this undated work, inv. 345, with the painting, not located, *In front of the Church of the Nativity of the Virgin in Moscow*, circa 1903–1904.

[17]One, a lead pencil drawing, the other made with lead and coloured pencils, these studies are part of the research for the realisation of *Moscow I*.

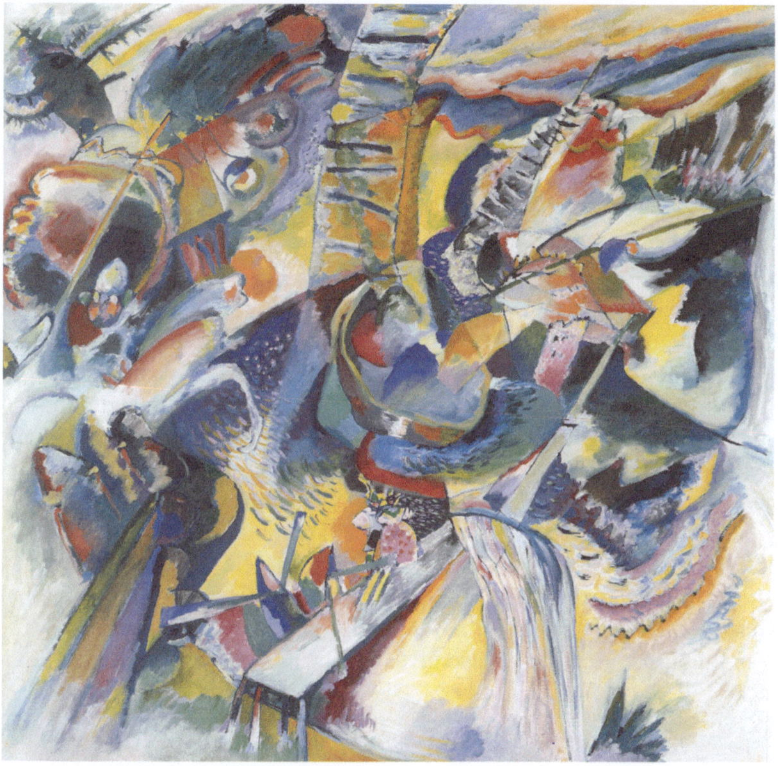

Wassily Kandinsky. *Improvisation Gorge*. 1914. Oil on canvas. 110 × 110 cm. The Städtische Galerie im Lenbachhaus, Munich

In all his production, Wassily Kandinsky seems to have reworked the imagination of his civilisation, which he felt close to, and which had intimately marked him. After years spent far from his country, he seems to reappropriate this environment, his cultural and personal background, to turn it into a pictorial world that finally places the subject in its midst. To this end, pictorial creation, iconic recognition, symbolic evocation, everything seems to be taken into account so that the viewer's perception participates in the work of art. Plastic and figurative elements operate in symbiosis for this purpose. However, nothing appears as being calculated in this painting. Rather, everything is organised according to that "organic connection" of which he had become aware (ibid., English transl.: p. 29) and which he had been pursuing since *Concerning the Spiritual in Art*.

Even in its topological structure, *Moscow I* is organised according to an organic relationship. The straight 'bridge' that supported the couple in *Improvisation Gorge* loses its geometric rigidity to harmonise with the biomorphic composition of the whole. Beyond the small open gate that somehow delays the introduction into the picture, the two personages stand in the very middle of the painting: perceptive centre in relation to the dimension of the canvas, but also spatial middle with respect to the planes and depths of the picture. The couple thus seems to lean forward towards the abyssal centre of the "basic plane" (cf. *supra*), where the space precisely deforms itself to host its centre of origin. From the impression of a diving perspective, such as a bird's-eye view, that dominates the very foreground, one perceives the gradual rise of these figures (which makes the point of view descend), then shifts to the impression of a slight low angle with respect to the urban crest at the top right, which finally unfolds in frontal perspective strongly inclined towards the left. The use of these bold points of view can be explained through the context, in the light of the exchange and possible mutual influences developed at that time in contact with the work of Aleksander Mikhailovich Rodchenko.[18] Nevertheless, Kandinsky appropriates this research in a personal way and combines several points of view to depict the dynamic complexity of this space. If the effects of these perceptual depths seem to lead to an upward oblique tension, a rotary movement immediately seizes it, and makes everything whirl all around. Not only does it attack the shapes of the objects and their location, but it also gives a twirling appearance to their lighting.

---

[18] As Nina Kandinsky pointed out, the young Rodchenko used to work in the "studio for art reproduction, a kind of graphic studio" that Wassily Kandinsky had installed on the sixth and last floor of his building near the Zubovskaya Square (Derouet and Boissel (eds.) 1984, p. 148, own translation). It is especially in photography that Rodchenko will then explore the effects of bold perspectives.

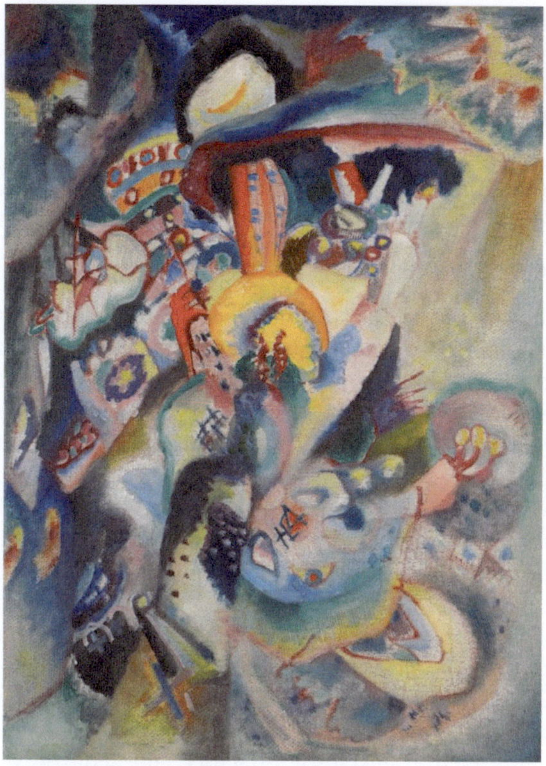

Wassily Kandinsky. *Moscow II*. 1916. Oil on canvas. 52.8 × 39 cm. Private collection

This complex dynamic, however, does not resolve itself in a simple spiral (see also the tension impressed on the rainbow). It seems to animate the basic plane with the same centrifugal (and, in return, centripetal) force that Kandinsky used to employ in due course in the morphogenetic act of pictorial creation. The referral instinctively goes to the photographic records, taken by Gabriele Münter, of the successive phases in the genesis of *Composition VII*, created by Kandinsky in 1913. Several works of art contemporary to *Moscow I* displays a structure organised according to this relation between centre and periphery, their perception referring to the probable creative movement: from the origin to the encompassing and, inversely, from the encompassing to the encompassed (see, among others, *(Untitled)* and *Naïve*, both 1916). Yet in *Moscow I*, this circular movement becomes more complex, due to the reintroduction of the semi-figurative, the perspectival depth, not to mention the upward topological tension.[19] The result is neither a kinetic dominance nor a sum adding up the perceived movements, but a complex dynamic that manifests both its logic and its organic contradictions. Kandinsky gives to the plastic means the power to

[19]In *Moscow II*, Kandinsky seems to shift away from this perspectival depth to focus more on plastic plane and the possible effects of optical depth generated by colour.

express all their tensions (compression, extension, condensation towards the centre and primordial emptiness, deformation, scattering, etc.). In the forming process, matter then regains all its cosmogenetic force, and art finds its "inner necessity"[20] (Kandinsky 1913, English transl.: p. 29):

> Painting is a thundering collision of different worlds intended to create the new world within and out of their strifes. This new world is the painting. Technically every work of art is created as the cosmos was—through catastrophes, which, from the chaotic roar of the instruments, finally form a symphony, which is called the music of the sphere. The creation of the work of art is the creation of the world. (Ibid., English transl. modified)

Moscow I is thus one of the artistic outcomes of a psychophysical theory which, while varying its name (psychophysics, psychophysiology, optico-psychology, physico-psychology), did not stop evolving in the thoughts of theoreticians and plasticians. It was Gustav Theodor Fechner, physicist turned to the philosophy of nature and to psychology, who formulated the term in 1850[21]; then the aestheticians gathered around the so-called pure visibility (Fiedler, Hildebrand, etc.[22]) adopted the concept. When Kandinsky is concerned with the "optico-psychological" value of painting, namely, how "colour is perceived optically, experienced psychically" (Kandinsky 1975, p. 45, own translation), it would be advisable to keep in mind, on the one hand, the visual and sound experiments of von Helmholtz and the psychophysiology of Wundt—reflecting a certain continuity of this tradition—, on the other hand, the contemporary research of the Gestalt theoreticians who, within the Berlin School (Köhler, Wertheimer, Koffka), renewed the psychophysical theory by a dynamic approach to the perception of form.[23] Similarly, when Kandinsky examines the kinaesthetic, tactile, gustatory, olfactory and acoustic sensations evoked by the vision of colours or, still, the dynamic and affective tensions conveyed by the geometric dimensions of point, line and plane (cf. Kandinsky 1975, pp. 46–59, 230 and 72–98), it would be appropriate not only to consider the aesthetic research on synaesthesia and on the synthesis of the arts but also the renewal that affected some art historians and theoreticians at that time as far as perception was concerned (e.g. Wölfflin 1928 on the anisotropy of the visual field; and Panofsky with his opposition between the symbolic construction of the Renaissance perspective and the

---

[20] Already in *Der Blaue Reiter* Wassily Kandinsky stressed the value of form as a means of expressing inner resonance ("*Ausdrucksmittel des inneren Klanges*"). Kandinsky, "Über die Formfrage", in: Kandinsky and Marc (eds.) (1911, new edition: pp. 74–100, esp. 76).

[21] Fechner briefly outlined what he called the "psychophysical basic law" in his *Zend-Avesta*, published in 1851, and further elaborated on his findings in *Elemente der Psychophysik* (*Elements of psychophysics*, 1860). Ellenberger (1970, p. 218).

[22] Konrad Fiedler argued in 1887 that it was impossible to distinguish psychic and physical processes as regards what he called the *expression* of form (cf. Fiedler 1887, French transl.: p. 73–74). In 1893, the sculptor and theorist Adolf von Hildebrand also expressed in terms of sympathy, activation of the direct feeling of the body, effects and appearance of form (cf. Hildebrand 1893, French transl.: pp. 117–145, esp. 137–138 and 122–126).

[23] Psychophysical ideas reappear, for example, in Wolfgang Kölher's essay *Gestalt Psychology* (cf. Kölher 1929, French transl.: p. 176). For an introduction to the Gestalt theory, cf. Guillame (1979).

"«reality»" of the "actual subjective optical impression" of the "psychophysiological space"; Panofsky 1927, English transl.: pp. 29–30). Anyhow, through various channels, an organic conception would assert itself and question—through perception—the ways in which the sense emerges from matter.

References

Caliandro, Stefania: *Moscou I* de Wassili Kandinsky (ou la perception psychophysique dans l'art). Visio, 8, 1–2, spring-summer, Quebec, pp. 147–158 (2004)

Comerlati, Doriana (ed.): Wassily Kandinsky tra Monaco e Mosca. 1896–1921. Exhibition catalogue. Mazzotta, Milan (2000)

Derouet, Christian; Boissel, Jessica (eds.): Kandinsky. Œuvres de Vassily Kandinsky (1866–1944). Exhibition catalogue. Centre Pompidou, Paris (1984)

Ellenberger, Henri F.: The Discovery of the Unconscious. The History and Evolution of Dynamic Psychiatry. Basic Books (Harper Collins Publishers Inc.), New York (1970)

Fiedler, Konrad: Über den Ursprung der künstlerischen Tätigkeit (1887) [French transl.: Sur l'origine de l'activité artistique. In: Salvini, R. (ed.) Pure visibilité et formalisme dans la critique d'art au début du XXe siècle. Klincksieck, Paris. pp. 70–100 (1988)]

Guillaume, Paul: La psychologie de la forme. Flammarion, Paris (1979)

Hildebrand, Adolf von: Das Problem der Form in der bildenden Kunst (1893) [French transl.: Le problème de la forme dans les arts plastiques. In: Salvini R. (ed.). Pure visibilité et formalisme dans la critique d'art au début du XXe siècle. Klincksieck, Paris, p. 117–145 (1988)]

Kandinsky, Wassily; Marc, Franz (eds.): Der Blaue Reiter. R. Piper & Co. Verlag, Munich, Zurich (1911). New edition: R. Piper & Co. Verlag, Munich, Zurich (1914)

Kandinsky, Wassily: Rückblicke. Berlin, Munich (1913). Russian edition revised by the author: Moscow (1918) [French transl.: Regards sur le passé 1913–18. In: Regards sur le passé et autres textes 1912–22. Hermann, Paris, pp. 83-141 (1974). Partial English transl.: Retrospect by Wassily Kandinsky. In: Rebay, H. (ed.). Kandinsky. the Solomon R. Guggenheim Foundation, New York, pp. 23–33 (1945)]

Kandinsky, Vassily: (On a method for working with synthetic art. Lecture at the Russian Academy of Artistic Sciences [RAKhN]). (1921) [French transl.: De la méthode de travail sur l'art synthétique. In: Derouet, C.; Boissel, J. (eds.) Kandinsky. Œuvres de Vassily Kandinsky (1866–1944). Exhibition catalogue, pp. 158–159. Centre Pompidou, Paris (1984)]

Kandinsky, Wassily: Punkt und Linie zu Fläche. Albert Langen, Munich (1926) [French transl.: Point—Ligne—Plan. Contribution à l'analyse des éléments picturaux. Denoël, Paris (1970)]

Kandinsky, Wassily: Abstrakte Malerei. Kronick van Hedenaagse Kunst en Kultuur, 6, Amsterdam (1935). New edition in: Bill, M. (ed.). Kandinsky. Essays über Kunst und Künstler. Verlag Gerd Hatje, Stuttgart, pp. 172–180 (1955) [French transl.: Peinture abstraite. In: Écrits complets, vol. II (La forme). Denoël-Gonthier, Paris (1970)]

Kandinsky, Wassily: Cours du Bauhaus. Denoël, Paris (1975)

Klee, Paul: Das bildnerische Denken, collected and edited by Jürg Spiller. Schwabe & Co, Verlag, Basel, Stuttgart (1956) [English transl.: Notebooks. Volume 1. The thinking eye. Lund Humphries, London (1961)]

Köhler, Wolfgang: Gestalt Psychology. Liveright, New York (1929) [French transl. of the edition revised by the author: Psychologie de la forme. Introduction à de nouveaux concepts en psychologie. Gallimard, Paris (1964)]

Panofsky, Erwin: Die Perspektive als 'symbolische Form'. Vorträge der Bibliothek Warburg 1924–1925. Leipzig, Berlin, pp. 258–330 (1927) [English transl.: Perspective as Symbolic Form. Urzone, New York (1991)]

Wölfflin, Heinrich: Über das Rechts und Links im Bilde. Münchner Jahrbuch der bildenden Kunst, series NF, 5, pp. 213–224 (1928). New edition in: Wölfflin, H. Gedanken zu Kunstgeschichte. Benno Schwabe & Co. Verlag, Basel (1946)]

Chapter 7
Eight Theses for (or against?) a Semiotics of Colour

Hard sciences (optics, physics, chemistry, neurology) and humanities (philosophy, art, anthropology, etc.) have never ceased to be concerned about colour. Countless theories have marked the thinking and imagining of colour, in which symbolic values and scientific methods cross, sometimes in a heterogeneous manner. Semiotics, for its part, has developed many reflections and analyses on the subject of colour, to such an extent that a semiotics of colour seems even to be set. I will propose a step back to rethink in what terms we can, or cannot, take in consideration a semiotics of colour.[1] I do not aim here at a semiotic analysis of one or more colours, even if I will refer to their use in art. Nor do I aim at a new theorisation of colour. But I would like to have a slight rearward movement to better position the reflection on colour in the semiotic field. To this end, I will draw a parallelism with the questions that Hubert Damisch posed in 1974 at the very foundation of a semiology of painting, as if, once the orientation had been given, semiologists and semioticians have gone beyond the verification of the bases on which their discourse and work rest, whereas these bases still show the uncertainties of the terrain involved. Without pretending to equal the eloquence of that inspiring author, this step back—going through some major problems raised by colour—will help us to grasp the *complexity* of the relationships at stake and will give us the impulse to reformulate, perhaps, the way to approach colour in future research.

1. What Semiotics of Colour?

Is there a semiotics of colour? Is colour semiotic and, if yes, in what sense it can be considered within a semiotics, given that, despite their divergent conceptions, the different visual approaches (Jean-Marie Floch, Groupe μ, Fernande Saint-Martin, etc.) seem at least to agree that colour is not, in itself, a semiotics? Is colour only

[1] This paper has been published in Spanish in the Mexican journal *Tópicos del Seminario* (Caliandro 2013).

S. Caliandro, *Morphodynamics in Aesthetics*, Lecture Notes in Morphogenesis,
https://doi.org/10.1007/978-3-030-29631-5_7

destined to be a semiotisable object, or more specifically to be semiotised by a semiotics that defines it within its own discourse, that is to say, by using the terms of the distinction established by Hjelmslev (1975, p. 9), would there be no *semiotics* of colour except to understand it as a *semiology* of colour? The question, thus posed, not only runs the risks of excluding from the semiotic fields the chromatic work which is that of painters and artists, or even in general the *logos* that determines creation and plastic shaping[2]; but also demands to delimit the conditions of enunciation in which the semiotic discourse is able to take on the analysis of colour. In other words, not only would there be solely a semiotics *applied* to colour—and yet the semiology cannot refrain from resorting to the plastic thinking that motivates art and from referring to artist's theorisations and spontaneous declarations—but the semiology must necessarily also be subject to the unavoidable constraints of a discourse that is second. In such a discourse, colour would have a fragmentary value in relation to the *system* of the work of art. It could only acquire a central value, although *approximate*, within a comparative or transversal examination, for example, of colour iconology in different times and cultures.

2. Describe the Colour

If one accepts to conceive the work of art as a "*system of signs*" (Damisch 1974, § 2, English transl.: p. 259) and if one recognises, as it is henceforth widely shared, that no minimal unit of sense is identifiable in art, one might believe that the above-mentioned questions concern as well colour as other constituent elements taken singularly (such as line, shape, material, light, topological position). None of these elements give rise, in itself, to a semiotics and it is only in the network of their reciprocal relations, that is, of their *articulation* each time very specific, that a semiotics can be conceived. Still, semiotic analysis can be carried out of one or another of these elements, treated as an iconological motif, the evolution and the signifying dimensions of which will stand out.[3] And yet it will be necessary to agree, in line with Ludwig Wittgenstein, on the particular difficulty to face in the matter of colour when one wants to provide a semiological description of it. In this respect, even the historian Michel Pastoureau, a renowned specialist in the study of colour, has ended one of his books with Wittgenstein's meagre statement on the difficulty of explaining the meaning of the words that relate to colour (cf. Pastoureau 2010a, p. 242). As the philosopher has detailed, we

[2]In contrast with this exclusion, let me refer please to the idea of a *logos of art* that I have elaborated from Paul Klee's work in my article "Sentir la complexité, dire l'inconsistant" in *Littérature*. The issue of this journal, edited by Denis Bertrand and Jean-Claude Coquet, was devoted to the theme: "Comment dire le sensible? Recherches sémiotiques" and invited to reflect, inter alia, on the distinction between *saying* and *describing*. Caliandro (2011) translated herein as Chap. 9.

[3]See, for instance, the much appreciated paper by Denis Bertrand, "Enthymème du visible: autour de la ligne", at the 2010 Congress of the IAVS in Venice, which, however provoked, in the audience's reactions, the question of knowing to what extent it could lead to a semiotics of the line. Cf. the proceedings of the Congress edited by Migliore (2011).

can easily distinguish and point at colours, but "we have either no idea at all of their use, or a very rough and to some extent false one" (Wittgenstein [1950–51] 1977, § I-68 p. 11e).[4]

If colour seems to be related to the hypoiconic notion of *image*, in the sense of Charles Sanders Peirce (cf. Peirce 1931–1958, 2.276–2.277)—a quality that can "*roughly*" be described (Wittgenstein [1950–51] 1977, § III-257 p. 50e)—perhaps it is because, as Pastoureau states, "colours are first of all concepts, ideas, intellectual categories. Then, they are words [...]. To speak of colours we are prisoners of these words. Finally—but only in the end—colours are matters, lights, perceptions, sensations" (Pastoureau 2010b, p. 232, own translation). Nonetheless, colours acquire meaning only insofar as they are matters, lights, perceptions and sensations. If they cannot be delimited with language, it is possible, however, to "intuitively" apprehend their effects—to take up the words of Itten (1961, English transl.: p. 7)—provided they are *com-prehended* with the other elements, *understood* within the complexity of the system in which they appear.

3. Colours: Matters, Lights, Perceptions and Sensations

That colours are perceptions and sensations is an evidence pertaining to common experience. From Peirce's viewpoint, bringing them closer to perception and sensations or, better, to feelings seems to underline the firstness of colour in relation to semiosis. Also artists have expressed themselves on the power of colour to affect psychic state and emotions, to the point that, in his project for a "science of painting", Wassily Kandinsky even evokes "coloured baths" to have an influence other than by the eye on the psychology of normal and pathological cases (Kandinsky 1921, own translation). As for the relation between colour and perception, it will suffice to recall, for example, Fernande Saint-Martin's approach, which defines the "*coloreme*" as "the basic unit of visual language", constituted by "the aggregate of visual variables perceived in the visual representation", where the coloreme is in fact conceived as "the zone of the visual linguistic field correlated to a centration of the eyes" (Saint-Martin 1987, English transl.: p. 5). More precisely, colour, one of the variables taken into account in her notion of coloreme,

> is the product of a perceptual phenomenon localized at the interfaces of matter which are in contact with air. [...] But, strictly speaking, color cannot be considered as a 'property' of matter itself. It consists rather of the spectral composition of the light reflected from an object, according to its specific structure of absorption and reflection of light rays. (Ibid., English transl.: p. 18)

This brings us back to Pastoureau's idea that colours are also matters and lights, even if the conjunction of these notions in semiotics has not always been obvious.

> In several language families, the etymology of the word that designates colour attests how this was first conceived and perceived as a matter, an envelope that covers beings and things.

---

[4]This phrase was deleted in the quote from Pastoureau that has just been cited here.

> This is particularly the case in Indo-European languages. The Latin word *color*, for example, from which the Italian, French, Spanish, Portuguese and English terms designating colour are derived, is related to the extended family of the verb *celare*, which means "to hide", "to wrap", "to conceal": colour is what hides, covers, dresses. It is a material reality, a film, a second skin or a second surface that conceals the body. The same idea is found in Greek: the word *khrôma*, "colour" derives from the word *khrôs*, "skin", "body surface". The same goes for the Germanic languages: the German term *Farbe*, to give only one example, comes from the common Germanic **farwa*, which means "form", "skin", "envelope". Other languages, not at all Indo-European, convey a similar idea: colour seems to be at the origin a matter, an envelope, a film.
>
> However, the lexicon is one thing and the theories of scientists or philosophers are others. Early in Europe colour ceased to be considered only as a matter to become, also and above all, a light; or rather a fraction of the light. (Pastoureau 2010a, pp. 237–238, own translation)

From Aristotle to Isaac Newton's prism experiments, as Pastoureau argues, colour was not only a material envelope, but also a physical phenomenon strictly related to light, as well as a coloured effect sensation whose perception was to be studied further in neurosciences (cf. ibid., pp. 238–239). Hence, one might wonder about the pertinence of the approach adopted by many semiotic analyses that attempt to distinguish the chromatic, eidetic, material, luminous and topological features of a work of art,[5] although, these elements co-determine each other interdependently. The artificiality of this distinction is especially striking as regards the relation of colour to light and matter, since the former arises from the phenomenal encounter of the latter two. If the colour can reveal the material that composes the work of art or hide it thanks to the chromaticism of a second skin, it can also vary according to the light beam that invests it. A semiotics of light[6], for instance, should then include a semiotics of the colour and the materials involved. By orientating ourselves towards a naturalisation of the perceptual structures, it is finally not surprising that even the remarks of a logician philosopher such as Wittgenstein invite us to articulate the reflection on colours in relation to a phenomenological apprehension:

> 232. When psychology speaks of appearance, it connects it with reality. But we can speak of appearance alone, or we connect appearance with appearance. […]
>
> 234. Psychology connects what is experienced with something physical, but we connect what is experienced with what is experienced. (Wittgenstein [1950–51] 1977, § III-232 and 234 pp. 47e–48e)

[5] See, in particular, Marsciani and Zinna (1991, p. 31), and Valenti (1991, p. 20). Actually, Algidras-Julien Greimas has been more cautious in his distinction of the categories, for example, the eidetic and the chromatic in painting. According to him, the distinction of these does not depend on the materiality of the signifier, but on their relational conception, which the beholder contributes to construct. Cf. Greimas (1984, Italian transl.: p. 43).

[6] Cf. for example Fontanille (1995).

4. Colour and Matter

Colour is matter (pigments and light) and, as soon as one touches matter, colour changes. Yves Klein, one of the artists most committed to the search for the immaterial in art, had to acknowledge this when, inspired by the indefinable of Delacroix (cf. Klein 2009, p. 27), he wanted to attain spirituality through colour. He thus tells of his struggle against any fixative binder, which inevitably alters the chromatic radiance of the pigment: as soon as the grains are ground and oil is mixed in, for instance, the colour is transformed (cf. ibid., p. 39). The entire history of art relates a constant search for pigments and materials from which colours result. As *De coloribus libellus* (Booklet on colours) by Antonio Telesio of Cosenza indicates, some names of colours derive from the matter from which they are extracted, the *anthrax* from coal, the *ferrugineus* (rust) from the oxidised iron, the *purpurereus* from purple, the *aureus* from gold and so forth for others (cf. Telesio 1549). This intrinsic relation of some colours with matter has not escaped Wittgenstein, for whom not all colour concepts are logically of the same kind: there is a logical difference, even before ontological, between, on the one hand, the concepts of 'colour of gold' or 'colour of silver' and, on the other, the concepts 'yellow' or 'grey'.[7] Not without some hesitation, Wittgenstein comes to propose a distinction:

> 254. There seem to be what we can call "colours of substances" and "colours of surfaces". (Wittgenstein [1950–51] 1977, § III-254 p. 50e)

To adopt the viewpoint of praxis,[8] as Wittgenstein suggests to do, and to perceive colour as the encounter of matter with light,[9] by recognising the different colour ranges that can be obtained by mixture of pigments or by light interaction, do not explain why some colours cannot be produced in a light beam. Besides the questionable existence of a white light—a difficulty that converges with the problem of the existence of absolute white and absolute black, even before they be called *colours*—the impossibility of perceiving a light as grey, brown or, finally, black, repeatedly attracts Wittgenstein's attention[10]: "There is no such thing as luminous grey" or, at

---

[7]Cf. Wittgenstein ([1950–51] 1977, § III-241 p. 48e, §§ III-54 and 52 p. 23e), as well as § I-33 (p. 7e) which links this topic to the relation of matter with light:

> 33. We speak of 'colour of gold' and do not mean yellow. "Gold-coloured" is the property of a surface that shines or glitters.

Wittgenstein considers, in addition, other metals (ibid., § III-258 p. 50e):

> 258. Think of the colours of polished silver, nickel, chrome, etc. or of the colour of a scratch in these metals.

[8]Cf. especially ibid., § III-317 pp. 58e–59e.

[9]Cf., for instance, ibid., § III-95 p. 28e, §§ III-171–174 pp. 40e–41e.

[10]Intuitively, one can understand the difficulty of perceiving a black light, since black is apprehended as a subtraction of light, while the difficulty remains noticeable in the case of a grey or brown light. Cf. ibid., § III-215 p. 46e and ff.

least, a light cannot be perceived as grey.[11] Itten seems to have circumvented the problem when, in describing "the blue mosaic walls" of the mausoleum of Gallia Placidia at Ravenna which bath "in an orange light, filtered through narrow windows of orange-tinted alabaster", he does not focus on the grey light effect that one would expect to see (Itten, 1961, English transl.: p. 9). While he has just explained the complementarity of orange and blue which, mixed, engender grey, Itten evades the idea of a grey light: "As the visitor moves about the shrine, he receives different quantities of light, which is alternately accented blue and orange, the walls reflecting these colors at ever-changing angles." (Ibid.) The topic is not further studied by the Groupe μ which merely indicates in parentheses the impossibility of obtaining brown, gold and silver colours from three lights—red, green and violet—whereas all other colours are possible (cf. Groupe μ 1992, p. 229). Should one not then wonder whether this perception of colours, produced by the encounter of light with matter, is rather subject to a semiosis that originates already in the psychophysical modalities of our sensory organs?

5. The Truth of a Colour

If the dichotomy matter colours / surface colours can reactivate referential-type thoughts—when, for instance, a *representation* evokes or simulates materials, such as marble or precious stones, and their properties, such as brilliance, opacity or transparency, in particular of a glass—if it can play on the use of certain materials to associate with their *presence* a different meaning—divine, spiritual or symbolic for gold,[12] the wealth of devotional intention for ultramarine lapis lazuli, etc.—this material and/or referential dimension does not indicate *in itself* the *value* that the relevant item assumes in the work of art. Already at the perceptual level, its chromatic appearance is strongly influenced by the surrounding colours. On the signifying plane, its semiotic value also depends on the arrangement and configuration of the entire work of art,[13] as evinced by Wittgenstein's paradoxical reasoning, reducing a painting to a puzzle:

> 60. Imagine a painting cut up into small, almost monochromatic bits which are then used as pieces in a jig-saw puzzle. Even when such a piece is not monochromatic it should not indicate any three-dimensional shape, but should appear as a flat colour-patch. Only together with the other pieces does it become a bit of blue sky, a shadow, a high-light, transparent or

[11] Ibid., § III-81 p. 27e. Nowadays, one could admit the existence of luminous grey, in particular, the reflective strips that are used in some garments to be visible at night. But to speak properly, it is not perception of a grey light.

[12] See, for example, the use of gold in Pre-Raphaelite painting or in Art Nouveau; cf. Benedetti et al. (2011).

[13] I have analysed how the use of certain materials, such as lead and straw, with their own colouration, activates a semi-symbolic *modus significandi* in the art of Anselm Kiefer. The signifying potential, however, is actualised in function of the semiotic relationships woven into the specific work of art. Cf. Caliandro (2009).

> opaque, etc. Do the individual pieces show us the *real colours* of the parts of the picture? (Wittgenstein [1950–51] 1977, § I-60 p. 10e)

In this sense, even the *truth* of a representation that seeks a referential anchorage in the natural world subscribes rather—with all that this implies in theory of meaning—to the *truth of painting* or, in general, of art.[14] As the logician philosopher writes, a blot of colour means a light, a cloth, a transparent glass or a precious material only in connection to its positioning with the other parts of the work of art.[15] Similarly, colour may mean "the property of a point in space" (ibid., § III-107 p. 30e), its distance in perception and its position in relation to the light source, which assigns it matt aspect, brightness or shadow effects. Empiric data in fact leads us to see something white behind something that is transparent and red, even though the former is red-painted; or to adapt our perception of colours in using tinted lenses (cf. ibid., § III-108 p. 30e, §§ III-171–174 pp. 40e–41e and §§ 7–8).

> 213. [...] And this is w[h]y I think it is worthless and of no use whatsoever for the understanding of painting to speak of characteristics of individual colours. When we do it, we are really only thinking of special uses. That green as the colour of a tablecloth has this, red that effect, does not allow us to draw any conclusions as to their effect in a picture. (Ibid., § III-213 p. 46e)

A colour is such only in a given environment, within which both its perceptual dimension and its signifying values are constructed, and even the sensations that a colour generates.

6. Perceive Colour, Comprehend the Sensitive

If one approaches the colour, not from the point of view of general categories, but in the specific articulation of its relations in a given set, where colour ultimately defines itself, at what moment can one say to perceive it with the greatest precision, to be able to delineate its manifestation? The question cannot be reduced to the distinction, equally problematic, between form and colour; it concerns the very definition of the specific form of the colour in its phenomenal appearance. This is not perceivable at very small scales, the appreciation of mixtures, streaks, marbled shades varying according to the distance[16]; some ways of using colour may, moreover, play into subliminal perception; but also, and most importantly, a colour zone is almost never grasped uniformly.

---

[14]Damisch's paper (1974) begins with "the deliberately ambiguous saying or utterance of Cézanne—to which Jacques Derrida will be sensitive—: «I owe you the truth in painting and I will tell it to you»" (Derrida quoting Damisch who quotes Cézanne; Derrida (1978b, English transl.: p. 2). Cf. also Derrida (1978a)).

[15]Cf. Wittgenstein ([1950–51] 1977, §§ III-76–77 pp. 26e–27e.)

[16]The pointillism (or chromoluminarism) of Georges Seurat particularly provides the experience of retinal synthesis of unmixed pigments.

It is not only a matter of referring to the abuse of language when we name colours, when we categorise them with simple denominations (the primaries first, then the generic: red, blue, yellow, green, purple, orange, brown, grey, etc.), whereas specific colours (scarlet, cinnabar, vermilion) or technical references (a precise number in the colour chart) are mentioned less frequently; but mainly to note the chromatic fluctuation within perception. The experience of this phenomenon is, however, at the centre of a lot of artistic research, including the well-known works of Mark Rothko and James Turrell.[17] By asking to place his large canvas *Vir Heroicus Sublimis* (approximately 2.4 × 5.4 m, 1950–1951) in the narrow room of a corridor, Bernett Newman intended to immerse the beholder's gaze into the sensitive variability of wide monochromatic areas. The excess of presence thus generated in the subjectivity of the vision led Jean-François Lyotard to reassess the notion of sublime in relation with the work of this artist (cf. Lyotard 1988a; 1988b). Yet, to what extent does the perception of chromatic variables and variations touch the meaning of the work of art? To what extent is a semiotic analysis able to grasp such nuances? Will it not be necessary to take into account all the parameters (luminous, material, eidetic, topological …) simultaneously to apprehend both the appearance and the meaning of a colour? Is it possible to pass from a properly plastic thinking or, better still, from an intelligibility through senses to some form of discursive (metadiscursive) expression concerning the sensitive? Perhaps it would be necessary to understand this *intelligibility through senses* as something already semiotic.

7. From Matter to the Colour Form: The Singularity of the Relationships

Paul Klee's plastic thoughts help us to consider how colour is inscribed in the set of dynamics at play in a work of art: the *weight* of a colour, its extent, its brightness and its degree of saturation are found to be variously significant in function of the topological location and the depth assigned, as well as in relation to the other elements set therein (cf. Klee 1977, Italian transl.: esp. pp. 197–214 and 424–429).

From this point of view and taking into account, the difficulty of categorising colours in a non-simplistic way, it seems not very interesting to define the specific articulation of a colour in a work of art by the way of *invariants* abstract from any phenomenal or iconological manifestation. In Hjelmslevian terms, if a *substance* can still be recognised as participating, through its various components (optical, physical chemical …), in the formation of a colour, it is more difficult to demarcate, even theoretically,[18] the shift from *matter* to *form*. Undoubtedly, an entire tradition of Western painting (from the Venetian Renaissance colourists to Delacroix, from

[17]Cf. Saint-Martin (ed.) (1994) on Mark Rothko, and Didi-Huberman (2001) on James Turrell.

[18]For his part, Wittgenstein observes the difficulty of a logical demarcation between the colour concepts related to the forms of colouration and the colour concepts related to chromatic matter, since these "colour concepts sometimes relate to substances (Snow is white), sometimes to surfaces (this table is brown), sometimes to the illumination (in the reddish evening light), sometimes to transparent bodies." Wittgenstein ([1950–51] 1977, §§ III-255 p. 50e).

Impressionists to Fauves, from Art Brut to Transavantgarde and Bad Painting …) has accustomed us to seeing colour as a matter of expression that would be, if not spontaneous, at least more direct and passionate than the elaboration of the form. Probably, the terms of matter and form recover here multiple meanings that blur the application of Hjelmslev's concepts, especially since these two terms are charged with a long philosophical legacy. However, it seems that no clear and peremptory distinction marks the transformation of a chromatic matter into the colour form, apart from the establishment of the above-mentioned relationships that the colour has with the other elements of the work of art. On the other hand, the materiality recognised to the colour, even after being set in a form, encourages understanding it as one of the elements in which the contiguity, intuited by Hjelmslev, of the sense (*purport*) with the amorphous matter appears. Certainly, a psychoanalytic perspective can further search the deep meanings of colour symbolism; of course, the so-called materiality of colour gives rise to the simulacrum function that colour can have (the second skin). Nevertheless, the subtle link that Wittgenstein sought between colour and matter, between the perception of the former and the effects of the latter, indeed encourages rethinking how colour contributes to the meaning of art, should it follow the morphological dynamics of the whole composition or, on the contrary, allow another meaning to emerge.

8. "see, you say"[19]

To conclude, let us return to the difference between *thinking* (plastically) by colour and *saying*, even *describing*, the colour. Because at a first level is the work proper to the artiste to use, arrange colours, and with that it is tied up the theory that emerges from the practice itself. It is in this first level—which, according to Klee (1973), we can call *creative thinking*—that plastic semiotics[20] are generated, producing, among others, the transformation of chromatic matter into colour forms. Here occurs the passage from the possible to the becoming, on which Marcel Duchamp sketches his concept of *inframince*: "The possibility of several tubes of colour becoming a Seurat is the concrete «explanation» of the possible as infra thin" (Duchamp 1999, pp. 22; English transl.: n. p.). Then, at a second level, the logos of art (the saying *of* art) gives way to a saying *about* art. This involves, for Duchamp, "a certain inopticity": "I mean the difference between speaking about red and looking at red".[21] The passage from seeing to the discursive enunciation marks, therefore, a further[22] detachment

[19]"*voir, dis-tu*": the expression quotes from the title of Hubert Damisch's article "Les voir, dis-tu; et les décrire". Damisch (1981).

[20]For plastic semiotics we mean, in a broad sense, all the semiotics concerning plastic and visual arts, regardless the medium or the degrees of figurativeness involved.

[21]Duchamp. À l'infinitif. "Boîte blanche". In: Sanouillet (ed.) (1994, p. 118 and, on the same page, note n. 3 written by Duchamp in 1965, added to this edition).

[22]The first step is, evidently, the passage from artistic making to seeing.

from the plastic thinking, with the risk of a paradoxical blindness in which the aim of the first-degree semiotics is lost. It is up to a scientific semiotics (metasemiotics) to *describe* art—and colour—being aware of the *inconsistency* of its own discourse (cf. Lyotard 1977; new edition: p. 66); such a discourse is necessary second in respect to the works of art, and also forced, as Wittgenstein has shown, in an inevitably approximate description of colour, the most frequently passing through rough and, in part, false representations. If a scientific approach to art cannot dispense with this tortuous path, it should also not lose itself in a rambling interpretation without fulfilling a semiotic apprehension of art. The role played in it by the understanding of the perceptual structures, as they are actualised in the work of art, becomes crucial in order to basically anchor the sensations and the feelings experienced in the work of art itself.

> Parodying Merleau-Ponty, it will be said that in treating painting along with the perceived one can only miss the semiotic level in it and, therefore, miss painting itself, to the extent that a truth is working to reveal itself there, which is not immediately dependant on the order of discourse, but which has relation, in the highest degree, to perception. (Damisch 1974, § 8; English transl.: p. 266)

Conclusion

Given the variety and the breadth of the points dealt herein, it is difficult to come to a conclusion trying to frame the issues raised by colour from a semiotic point of view. Abandoned the rather fragile hypothesis that colour can be, in itself, a semiotics, the problem remains to know how colour can become a semiotisable object or, better still, at what stage it joins up a semiotic level. Taking in consideration the limits faced by colour description, the significance of its phenomenological definition in perception, and the determination of its value with respect to the relationships it establishes in a given environment, I have gone in-depth on the specific relationship of colour with matter and, in a Hjelmslevian perspective, I have proposed to identify the semiotic definition of colour in its transition from the chromatic matter to form, i.e. in the singularity that colour acquires through its relationships with other elements and tensions in the work of art. In this sense, it seems fair to find a primary semiotic conception in the perceptual dynamics that colour engenders, especially within a *plastic thinking* proper to art creation. It seems, henceforth essential to reassess this first level of intelligibility of the sensitive, to avoid certain perilous twists that language encounters when it claims to express colour.

References

Benedetti, Maria Teresa; Frezzotti, Stefania; Upstone, Robert (eds.): Dante Gabriel Rossetti, Edward Burnes-Jones e il mito dell'Italia nell'Inghilterra Vittoriana. Exhibition catalogue. Mondadori Electa, Milan (2011)

Caliandro, Stefania: O semi-simbólico na arte. Estudos Semióticos, 5, 1, pp. 1-8, http://www.revistas.usp.br/esse/article/view/49221 (2009)

Caliandro, Stefania: Sentir la complexité, dire l'inconsistant. Littérature, 163, pp. 20–31 (2011)

Caliandro, Stefania: Ocho tesis a favor (¿o en contra?) de uma semiótica del color. Tópicos del Seminario. Revista de Semiótica, 28, 2, pp. 21–38 (2013). Also at: http://www.topicosdelseminario.buap.mx/index.php/topsem/article/view/67/61

Damisch, Hubert: Huit thèses pour (ou contre?) une sémiologie de la peinture. Text of the general report presented to the first congress of the international association for semiotic studies at Milan, 2–6 June 1974; published In: Chatman, S., Eco, U., Klinkenberg, J.-M. (eds.) A Semiotic Landscape/Panorama sémiotique, Proceeding of the first Congress of the International Association for Semiotic Studies in Milan in June 1974. Mouton, Hague, Paris, New York, pp. 128–136 (1979). New edition: Macula, 2, pp. 17–23 (1977) [English translation: Eight Theses For (or Against?) A Semiology of Painting. Oxford Art Journal, 25 May 2005, pp. 257–267 (2005)]

Damisch, Hubert: Les voir, dis-tu; et les décrire. Versus. Quaderni di studi semiotici, 29, May-Aug, pp. 33–57 (1981)

Derrida, Jacques: Restitutions de la vérité en pointure. Macula, 3–4, pp. 11–37 (1978a)

Derrida, Jacques: La vérité en peinture. Flammarion, Paris (1978b) [English transl.: The truth in painting. The University of Chicago Press, Chicago, London (1987)]

Didi-Huberman, Georges: L'homme qui marchait dans la couleur. Les Éditions de Minuit, Paris (2001)

Duchamp, Marcel: Inframince. In: Notes (1st ed. 1980) Flammarion, Paris, pp. 19–36 (1999) [English transl.: Note 1. In: Notes. G.K. Hall, Boston, n. p. (1983)]

Fontanille, Jacques: Sémiotique du visible. Des mondes de lumière. Presses Universitaires de France, Paris (1995)

Greimas, Algidras-Julien: Sémiotique figurative et sémiotique plastique. Actes Sémiotiques. Documents, 60, Paris (1984) [Italian transl.: Semiotica plastica e semiotica figurative. In: Corrain, L., Valenti, M. (eds.) Leggere l'opera d'arte. Dal figurativo all'astratto. Esculapio, Bologna, pp. 33–51 (1991)]

Groupe, μ.: Traité du signe visuel. Pour rhétorique de l'image. Seuil, Paris (1992)

Hjelmslev, Louis: Résumé of a Theory of Language (Edited and translated in English by F. J. Whitfield). The University of Wisconsin Press, Wisconsin (1975)

Itten, Johannes: Kunst der Farbe. Otto Maier Verlag, Ravensburg (1961) [English abridged edition and transl.: The elements of color. A treatise on the color system of Johannes Itten based on his book *The art of color* edited and with a foreword and evaluation by Faber Birren. Van Nostrand Reinhold Company, New York (1970)]

Kandinsky, Vassily: (On a method for working with synthetic art. Lecture at the Russian Academy of Artistic Sciences [RAKhN]) (1921) [French transl.: De la méthode de travail sur l'art synthétique. In: Derouet, C., Boissel, J. (eds.) Kandinsky. Œuvres de Vassily Kandinsky (1866–1944). Exhibition catalogue. Centre Pompidou, Paris, pp. 158–159 (1984)]

Klein, Yves: Verso l'immateriale dell'arte. Con scritti inediti. Editions O barra O, Milan (2009)

Klee, Paul: Das bildnerische Denken. Papers collected and edited by Jürg Spiller. Schwabe & Cie Verlag, Basel, Stuttgart (1973) [French transl.: Écrits sur l'art/1. La pensée créatrice. Dessain & Tolra, Paris (1973)]

Klee, Paul: Unendliche Naturgeschichte. Papers collected and edited by Jürg Spiller. Basel, Stuttgart: Schwabe & Cie Verlag (1977) [Italian transl.: Teoria della forma e della figurazione. vol. 2: Storia naturale infinita: assetto di principio dei mezzi figurativi in connessione con lo studio della natura, e costruttivi mezzi di composizione. Feltrinelli, Milan (1970)]

Lyotard, Jean-François: Les Transformateurs Duchamp. Galilée, Paris (1977). New bilingual edition: Les Transformateurs Duchamp. Duchamp's TRANS/formers. Van Looveren & Princen, Bruxelles (2010)]
Lyotard, Jean-François: L'instant, Newman. In: L'inhumain. Causeries sur le temps. Galilée, Paris, pp. 89–99 (1988a)
Lyotard, Jean-François: Le sublime et l'avant-garde. In: L'inhumain. Causeries sur le temps. Galilée, Paris, pp. 101–118 (1988b)
Marsciani, Francesco; Zinna, Alessandro: Elementi di semiotica generativa. Processi e sistemi di significazione. Esculapio, Bologna (1991)
Migliore, Tiziana: Retorica del Visibile. Strategie dell'immagine tra significazione e comunicazione, vol. 1. Aracne, Rome (2011)
Pastoureau, Michel: Qu'est-ce que la couleur? Les couleurs de nos souvenirs. Seuil, Paris, pp. 237–242 (2010a)
Pastoureau, Michel: Parler des couleurs sans les montrer. Les couleurs de nos souvenirs. Seuil, Paris, pp. 232–235 (2010b)
Peirce, Charles Sanders: Collected Papers of Charles Sanders Peirce. In: Hartshone, C., Weiss, P., Burks, A. (eds.) 8 volumes. Harvard University Press, Cambrige, MA (1931–1958) (As a matter of convention, references to this publication give the volume number, followed by a full stop, and the indication of the relevant paragraph number)
Saint-Martin, Fernande: Sémiologie du langage visuel. Presses de l'Université du Québec, Sainte-Foy (Quebec) (1987). New edition: 1994 [English transl.: Semiotics of Visual Language, Indiana University Press, Bloomington and Indianapolis (1990)]
Saint-Martin, Fernande (ed.): Nouveaux Actes Sémiotiques, VI, 34-35-36, Approaches sémiotiques sur Rothko. Pulim, Limoges (1994)
Sanouillet, Michel (ed.): Duchamp du signe. Écrits. Flammarion, Paris (1994)
Telesio, Antonio [Antonius Thylesius de Cosentinus]: De coloribus libellus. [1st edn, 1528] (1549) [French transl.: Petit traité des couleurs latines. Éditions Estienne, Paris (2010)]
Valenti, Mario: Introduzione. In: Corrain, L., Valenti, M. (eds.) Leggere l'opera d'arte. Dal figurativo all'astratto. Esculapio, Bologna, pp. 7–30 (1991)
Wittgenstein, Ludwig: Remarks on Colour [Bemerkungen Über Die Farben]. Bilingual (German-English) edition by G.E.M. Anscombe. University of California Press, Berkeley, Los Angeles ([1950–51] 1977)

Chapter 8
Art as Crystallisation of Life. Considerations on a Work of Art by Edvard Munch (*From Nordstrand*, 1892)

I felt our love lay upon the earth like a pile of ashes.
(E. Munch in Tøjner (ed.) 2001, p. 126.)

If it is possible to talk about art theory (and art theories), it is because the theoretical object is in art. The theoretical object emerges from the work of art, in its conception as well as in its contemplation. And a theory acquires meaning only by realising itself in the artwork.[1] I would like to focus here on the *singularity* of a work of art, proceed less in the spirit of a totalising view, as it occurs when studying an artistic genre or theme or, still, an author's style, and insist rather on the appreciation of a work of art, comprehend its specific character, whilst relating it to its context. To that extent, Hubert Damisch's approach is inspiring insofar as his article "Narcisse Baroque?" lays stress to unveil the *folds* of a Caravage's painting, by pointing out aspects that arise from morphogenesis, i.e. *figurability* of forms in the representation (Damisch 1996).

This morphogenetic perspective, in my opinion, allows a better understanding of the relation between theory and artistic practice, especially whenever one is not ahead of the other, or vice versa, but, most frequently, the two coexist. In this respect, the work of art becomes a theoretical object, in a double sense: as object of a theory, and object that has the theory in itself. The genetic aspect, stressed by some approaches[2] reacting, on the one hand, to an increasingly decontextualized use of art and aiming, on the other hand, to reconsider the historical moment in which works of art participate, becomes interesting insofar as it highlights features related to genesis that contribute to apprehending the meaning of the work of art. By this I mean both the morphological traits which are part of the creative process and displayed in its manifestation—such as repainting, erasures, variants—and all the iconological,

[1] A first version of this paper has been presented in the workshop "Theoretical object «art» II", organised by Hubert Damisch and Omar Calabrese, in Urbino on 23 July 2006 and published in the *Bulletin de la Société française d'esthétique* (Caliandro 2007). I owe to the Munch Museum staff, in particular the kind assistance of Karen Elizabeth Lerheim and the art historian and senior curator Gerd Woll, relevant information provided to improve this paper.

[2] See the academic journal *Genesis*, in particular the issue n. 24, 2004, edited by Ségolène Le Men.

S. Caliandro, *Morphodynamics in Aesthetics*, Lecture Notes in Morphogenesis,
https://doi.org/10.1007/978-3-030-29631-5_8

stylistic and/or compositional filiations, through which the work of art interacts with the material imagination of its time. Concerning the latter point, the relation between art and artist's life is also to be considered, and the relevance of this relation can be examined with regard to the subjects painted and the aesthetic modalities of their realisation.

This relation between art and life becomes particularly important in the case of those artists who, like Munch, have sought to transmit and let transmit pathos through their art. However, I will not draw attention to the details of the author's trials and tribulations, since they help only marginally to understand the value of his art. I will therefore leave aside a good deal of *literature* about the artist, which, as criticism has remarked, goes far beyond the pictorial impact of his work.[3] Instead, I would like to grasp some aspects of his aesthetic from the analysis of the canvas *From Nordstrand*. My purpose is also to try to give voice to the idea of *crystallisation* formulated by the painter in his writings. According to this idea, art is a crystal, in the sense that it crystallises living forms, viz. it proceeds from crystallisation of life.

> An artwork is a crystal. A crystal has a soul and a mind, and the artwork must also have these.
>
> It is not enough for an artwork to have correct angles and lines.[4]

And also:

> The primeval light is everywhere and the light is where life is – and everything is movement and light.
>
> The crystals are born and are formed as a child in its mother's womb – and even in the hardest stone, the flame of life burns.
>
> Death is the beginning of a new life.
>
> Crystallisation.
>
> Death is the beginning of life.[5]

Furthermore, Munch explains in what terms "Art is the opposite of nature":

> An artwork comes directly from man's inner being.
>
> Art is an image, translated into form, which comes into existence through man's nerves. Heart. Brain. Eye.
>
> Art is man's need for crystallisation.[6]

Art thus derives from the "desire for crystallisation",[7] is this desire itself. It inspires itself from nature, "*which nourishes art*":

> Nature is not only that which is visible to the eye. It is also the inner images of the mind. The images upon the reverse of the eye.[8]

---

[3]Cf. Bjerke (2005, p. 21). The author mentions in particular Clement Greenberg's reaction.

[4]Munch in Tøjner (ed.) (2001, p. 100), ref. N 63, Ekely, Skøien, 1929.

[5]Ibid., p. 104, ref. T 2704.

[6]Ibid., p. 131, ref. N 57, Warnemünde, 1907–1908.

[7]Ibid., p. 136, ref. T 2742 *Music and Painting*.

[8]Ibid., p. 131, ref. N 57, Warnemünde, 1907–1908.

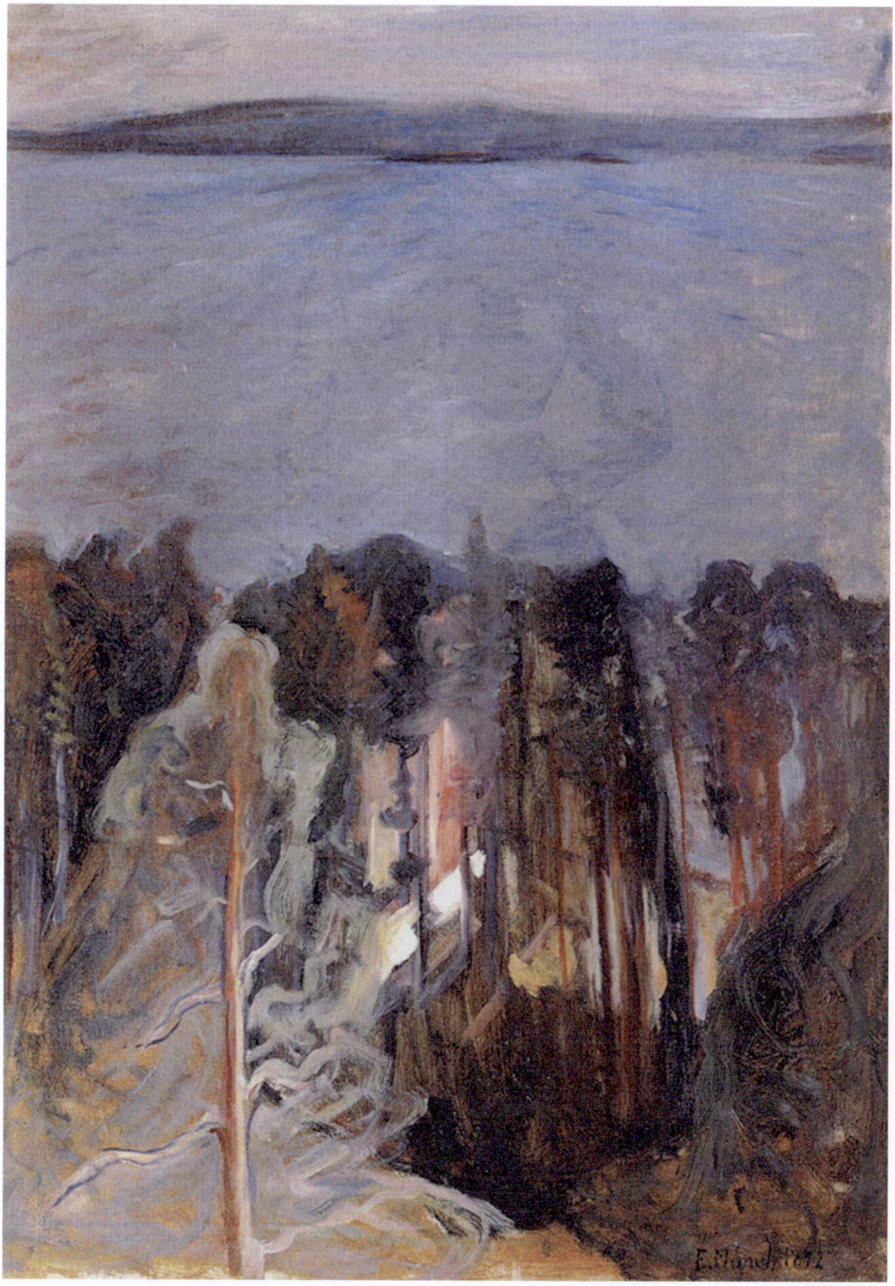

Edvard Munch. *Fra Nordstrand—From Nordstrand*. 1892. Oil on panel. 99 × 67 cm. Private collection

But how does a landscape, namely *From Nordstrand*, constitute a form of crystallization of life? How can it transmit the feelings of people who live, suffer, love, according to the saying repeatedly borrowed from the painter's pen by monographs on Munch? Here is what the canvas allows to see, at least at first glance. A forest occupies about the lower half of the picture, that is to say the core of the representation wherein a tree stands out in the very foreground on the left. In the upper part of the panel, the remaining landscape draws an expanse of water, separated from the sky by small reliefs. Criticism was willing to identify, by the place reference provided in the title, a fiord and three small islands to be seen at the bottom of the hills, on the bank opposite to the beholder's viewpoint. More delicate and questionable seems to me the recognition of a house or building inside the wood.[9] On the canvas, at the bottom right position, as traditionally for this purpose, the painter put his signature and the date (1892).

The pictorial treatment is quite different in these two parts of the painting. Both are worked with thin layers of oil colour, which are almost fluid in some places—such as the waterline above the forest, the salient tree and the vegetation on the right in the foreground—and leave, in several points, the impression of *non finito* to emerge, which the artist did not disdain (cf. Munch in Tøjner (ed.) 2001, p. 145, ref. June 1930). However, the facture is almost homogenous and uniform in the bluish chromaticism of the upper part, while the lower part, treated more with brown and warm colours, is constructed with unequal, diversely oriented strokes, placing close to each other strident tonalities, clashing together. Between the two parts an effect of strong tension results, which increases toward the centre of the wood, that is to say the centre of the picture, where discordant layers clearly oppose the serene expanse of the water and the background. Here, among what one perceives as flows and liquid infiltrations of rash colours, a crucial point of the representation takes shape, like a crackling fire in the middle of the forest.

The idea of a fire outlined inside the vegetation seems more likely than the hypothesis, suggested by the catalogue of an exhibition in Rome, discerning a construction in the wood. This perceptive possibility would find some support, not just in the dubious subtitle added to this painting in the same catalogue: *Burning Desire*,[10] but in Munch's writings, where the symbolic value of fire is evoked both to indicate the desire for reproduction and to mark the passion leading to the burning, to the destruction of life.

> The power of reproduction began – first as an unclear dream – incomprehensible desire – and finally complete desire. And it mated with another individual – bringing to life more fuel for the fire.

---

[9]See the painting description in the abovementioned catalogue: Bjerke and Bonito Oliva (eds.) (2005, p. 102).

[10]This exhibition catalogue reproduces the painting under the title: *Paesaggio a Ljan. Desiderio ardente* (literally, *Landscape in Ljan. Burning Desire*); cf. Bjerke and Bonito Oliva (eds.) (2005, p. 102). However, there is no mention of such title and subtitle in the catalogue raisonné edited at the Munch Museum, which gives the simple Norwegian title *Fra Nordstrand* (English transl.: *From Nordstrand*); cf. Woll (ed.) (2008). Although it must be reminded that Munch often left his friends or promoters give titles to his works of art, in an email to me Gerd Woll has advised that "Burning desire" has never been part of the title of this painting in any exhibitions or other documentation during Munch's lifetime.

> And I saw how an individual was born in those places where the masses had huddled together – another individual was born upon a freer earth – and both individuals had the same desire for destruction – for the combustion of life. (Munch in Tøjner (ed.) 2001, p. 104)

We feel indeed this landscape being charged with symbolic, passionate values that are specific to Munch's art production, embodying emblematically the attitude of the symbolist movement of that time. The rather vague, imprecise and almost sketchy definition of the landscape corroborates the artist's determination to paint landscapes from nature only through the vision that one had of them. The landscape is filtered, re-elaborated by an introspective approach, so that the first impression received by the artist is somehow transmitted through the painting.

> I do not paint what I see. But what I saw. (Ibid., p. 131, ref. Kristiania, 1890)

And Munch specifies in his late writings:

> It is better to paint a good, unfinished painting than finish a bad one.
>
> Many believe that a painting is finished when as many details as possible have been completed.
>
> A single line can be a finished work of art.
>
> A painting must be done with a sense of purpose and feeling. (Ibid., p. 145, ref. June 1930)

At that time, the idealistic movement advocated the introspection and the use of the inner imagination for the elaborations of contents.[11] At the same period, Paul Gauguin advised to adopt an internal vision to paint nature, and Émile Bernard encouraged the development of representation from *memory*. A similar conception appears in Munch's theoretical and plastic work.

> The truth of the matter is that one sees with different eyes at different times. One sees differently in the morning than in the evening.
>
> The way one sees is also dependent upon one's emotional state of mind. (Munch in Tøjner (ed.) 2001, p. 145, Ref. T 2761)

As he notes in his writings, even in spite of the resemblance to nature, Munch strives to give back in his landscape paintings the state of mind or the emotion experienced at the moment of the contemplation (cf. ibid., p. 64, ref. N 73). The importance of the impression received by the subject leads him to work on the particular imagery involved in the painted landscape. I have cited above how nature, for Munch, is also constituted by "the inner images of the mind", the singular images visible "upon the reverse of the eye". By confronting *From Nordstrand* with another of his paintings of the same period, *Pine forest*,[12] it can be seen how the latter focuses on the perceptual moment, in agreement with postimpressionist research (in particular, from Monet to Seurat), sharing this point with some achievements of Les Nabis (Bonnard, Vuillard) and even with several paintings of Klimt. Like Gustave Klimt's *Beech Grove I* (1902), *Pine forest* explores chromatic and luminous vibrations by juxtaposing small touches of paint, to exalt the abundance of ocular vision in a swarming surplus, profuse with

[11] Cf. Bjerke (2005, p. 25).

[12] *Pine forest* is signed, without date; the catalogue of the exhibition in Rome however inserts it in the production of the years 1890–1891 and reproduces it with the title: *Paesaggio con boschi al sole* (*Landscape with wood in the sun*). Cf. Bjerke and Bonito Oliva (eds.) (2005, p. 104). A few online sites refer to this canvas as *A sunbeam*.

optical effects. Very different is the treatment of the subject in *From Nordstrand*, in which the representation is, as it were, thicker, more dense and charged with dark tensions.

Without necessarily wanting to suggest a direct influence, this painting can intuitively be likened to Paul Cézanne's *The Hanged Man's House, Auvers-sur-Oise* (circa 1873), for the tragic aspect held by the image. Like the perturbing spatiality depicted by Cézanne, *From Nordstrand* places us in front of the threshold of the representation; the tree apart in the foreground on the right seems at once to determine and emphasise the beholder's top-down view of the landscape and the forest. Below the marginal terrain wherein the viewer is thus located, opens a precipice surprisingly similar, in its forms, to the declivity of *The Hanged Man's House*. In this way, Munch's painting makes the gaze dizzy and leads it to the gloomy part of the picture, wherein dark blacks increase the contrast with the rest of the landscape.

By daring an audacious, anachronistic comparison, since distant in modern art evolution, a certain relation could be found with the *Empire of Lights* (1954) of René Magritte, not least for the opposition of the two luminous registers, at the top and the bottom, and for the sense of mystery which is triggered by the piercing lighting in the heart of the dark. In *From Nordstrand* this radiance provokes major tensions in perception, as much for its luminosity, its pictorial treatment, the gesture inscribed in its facture, as for the interpretation that can be given as regards the representation. Indeed, where does the beholder's gaze dive? What stands beyond the precipice?

Increased attention leads to see what escaped me at first glance during the contemplation of the painting at the exhibition. If the size of the canvas[13] scatters, in some ways, the look inside the painted panorama, its reproduction in the catalogue renders the gestalt of a silhouette of woman, appearing beneath the painting of the landscape. Pentimento of a discarded representation, the traces of this picture were nevertheless not wiped away in the final figuration; on the contrary, it seems that, by the use of diluted layers and the black hatching delimiting its contours within the forest, the painter took care that the woman image does not disappear entirely. The painting shows to have been deliberately completed as if it were to display a moment of its genetic transformation in its ultimate realisation. The shape of the woman, whose forms, the waist and the dress's outline can easily be guessed, is not just given in the painting's stratification but appears through the landscape. Her presence is perceived in the nature. She stands upright exactly on the bottom of the precipice toward which the eye plunges.

More than the evocation of the feminine essence of nature, this figure in the substratum seems to add a poignant connotation to the configuration of *From Nordstrand*. Its positioning in relation to the final composition of the picture is not left to chance. The coincidence of the zone charged with major perceptual tensions and the lower part of the woman's dress raises the sexual meanings that the parallel with Munch's texts aroused. What has been identified as a possible fire burning in the dark background of the forest imposes itself here within the imprecise but remarkable contours of the feminine skirt. Indeed, in the lower part of this underlying figure opens the precipice of a burning desire. In this very place is consumed the desire for reproduction and destruction, the combustion of life expressed in Munch's aesthetic.

---

[13]The dimensions of *From Nordstrand* are 99 × 67 cm.

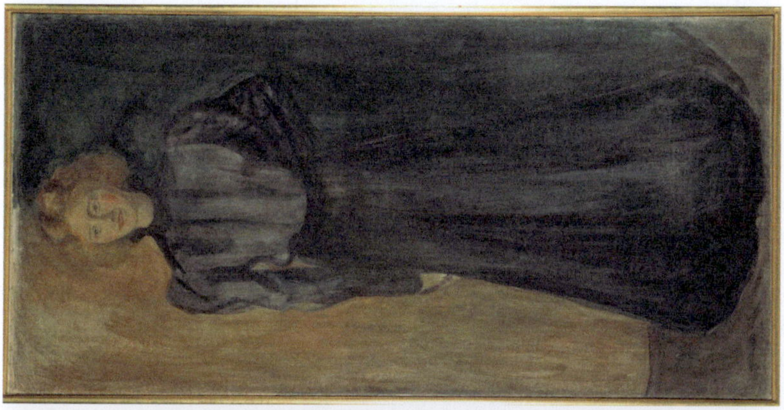

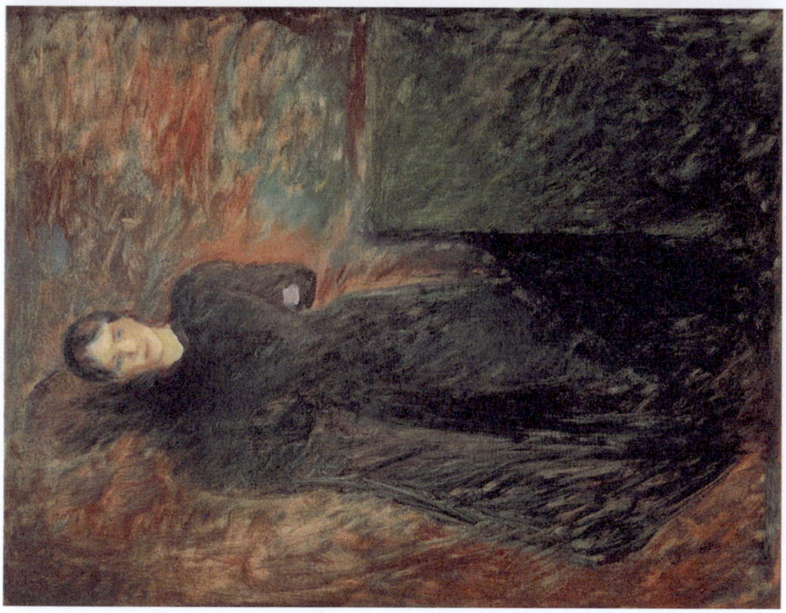

Edvard Munch. *Sortklædt dame—Lady in black*. 1891. Oil on canvas. 73.4 × 59.4 cm. Statens Museum for Kunst, Copenhagen. Edvard Munch. *Tulla Larsen*. 1898–1899. Oil on canvas. 119.5 × 61 cm. Munch Museum, Oslo. *Photo* © Munchmuseet

We thus manage to circumscribe some traits of the sensuality that this work of art emanates. Albeit the figure adopts here a still pose, as frequently occurs in his female portraits—e.g. the hieratic posture of *Lady in black* (1891) with her hands behind her back—the subliminal presence of the woman in *From Nordstrand* triggers a power equal in strength to the *Madonnas* realised by the artist. In one of the *Madonna* series, for instance,[14] the woman exposes her naked body, like a new erotic and almost heretical icon. With a red halo crowning her head, she pushes the left arm back, letting even suppose a silhouette of a man behind her. Although the hair encompasses the entire figuration in a fatal grip, as in Munch's iconological production concerning the vampire woman,[15] her head, lifted up, deformed into a few expressive points in an abandon suggesting pleasure—the *Madonna* series bears also the title *Woman who loves*—the head, as I was saying, is, in a disguised way, decapitated from her body by a bloody shadow.

A troubling sensuality also transpires in *From Nordstrand*, yet with other aesthetic modalities. Whilst in *Madonna* the sexual connotation is vividly displayed in the content of the representation, it almost creeps in a stealthy manner into the figuration of the landscape. Not only does the woman not exhibit her nakedness here, but she also tends to disappear in the originary background of the picture. Her perception is indeed given in a more subliminal than deliberate way. She fades like an archetypal figure to whom, despite the sought erasure, the landscape painting continues to hold. The superimposed brushstrokes and layers, while recalling the latent presence of the womanly figure, seek to destroy it. Better, they annihilate this presence by the superposition of plastic tensions which nevertheless continue to restore its potential, qua figurable. Interacting with their morphogenesis, the pictorial means allude to it to symbolise, all at once, the desire for combustion that Munch feels in his relation to woman, the act of reproduction-creation which concerns both nature and art, the moment of beheading-destruction of life represented by the blazing fire and by the extent of the landscape cutting off the figure.

This pictorial stratification *making* image entails thus the aesthetic of an art understood as crystallisation of life. By playing on ambiguity or, better, on the signifying and symbolic polyvalence of the plastic figurability, Munch implements his conception of an art *freezing* the living forms, crystallising for ever the potential fuel of life.

The artist mentions on many occasions the passion that his works of art would like to carry.

> The painter gives these paintings his dearest possessions – his soul – his sorrow – his joy – his heart's blood.
>
> He gives these things to humanity, not to the object. One would expect these paintings to take a stronger grip upon people – first a few, then several more, then everyone. (Munch in Tøjner (ed.) 2001, p. 79, ref. Nice, 2 January 1891 Nice)

---

[14]Edvard Munch, *Madonna*, oil on canvas, Nelson Blitz jr. & Catherine Woodard Collection.

[15]See in particular Munch's lithography *Vampire I* (1895) or his oil called *Vampire in the forest*, undated. The catalogue of the exhibition in Rome refers to the latter as *Donna che bacia un uomo sulla nuca* (*Woman kissing the neck of a man*), following the title given to the canvas at its first exhibition in 1918. Cf. Bjerke and Bonito Oliva (eds.) (2005, p. 176).

To properly transmit lived impressions, sensations and emotional states *through* his works of art, he also emphasises the importance of rendering the *distorted* perception that one has had in these moments.

> When one is out on a drinking spree, one sees things differently. Drawings become hazy, and everything seems more chaotic. It is a well-known fact that one sees things in a strange way.
>
> It seems obvious that one must then draw or paint in a strange way. If one sees double, one must, for example, draw two noses.
>
> And if one sees a crooked glass, one must paint it crooked.
>
> The same thing applies if one wishes to convey that which one has felt in an erotic moment when one is still heated and warm from love. Such a moment represents a motif that cannot be painted exactly as one sees it at that very moment – one must wait until one has cooled down. It is acceptable that the first image one has seen is quite different from the second. One experiences things differently when one is warm than when one is cold.
>
> It is this fact, and this alone, that gives art a deeper meaning. It is the human aspect – life, which one must try to convey. Not dead nature (still life). (Ibid., p. 145, ref. T 2761)

Stratification enables to give form, in this painting, to the reconstruction of an impression that can only be experienced in time. It plays on the superimposition of representations[16] and overlaps differentiated moments in the same space. The significance engendered by this picture conceived through layering, however, is not given by the sum of the meanings generated by each stratum considered in itself. It inevitably increases in their interaction and, moreover, in their interference. In the light of the quoted passage, actually written much later,[17] the hypothesis could be advanced, for example, that by this stratification of images *From Nordstrand* manages to condense the durative extent of the *desire for combustion*, evoked in his writings, to bring together different times, as for the object of desire (the feminine figure), the act of reproduction-destruction (the fire sketched out) and the pleasure of the contemplation (the landscape breadth). By the interference of these three moments, other meanings might be added such as, to put forward one, the value of the water depicted in its peaceful aspect, which overflows and comes to *cool* the perceptive tensions produced by the vegetation and the fire. However, what matters to me is not so much to provide this or that interpretative suggestion, as to understand how the artist delved into the plastic means to be able to open the work of art to such a range of particular paths in the perception of the imaginary. In fact, as he stated, we are less involved in what we see than in the images upon the reverse of the eye.

[16] Dario Gamboni (2002) studies other types of "potential images" in the representation.

[17] It should be noted that the text seems to have been written around the 1930s. In addition, it does not make any direct reference to *From Nordstrand*.

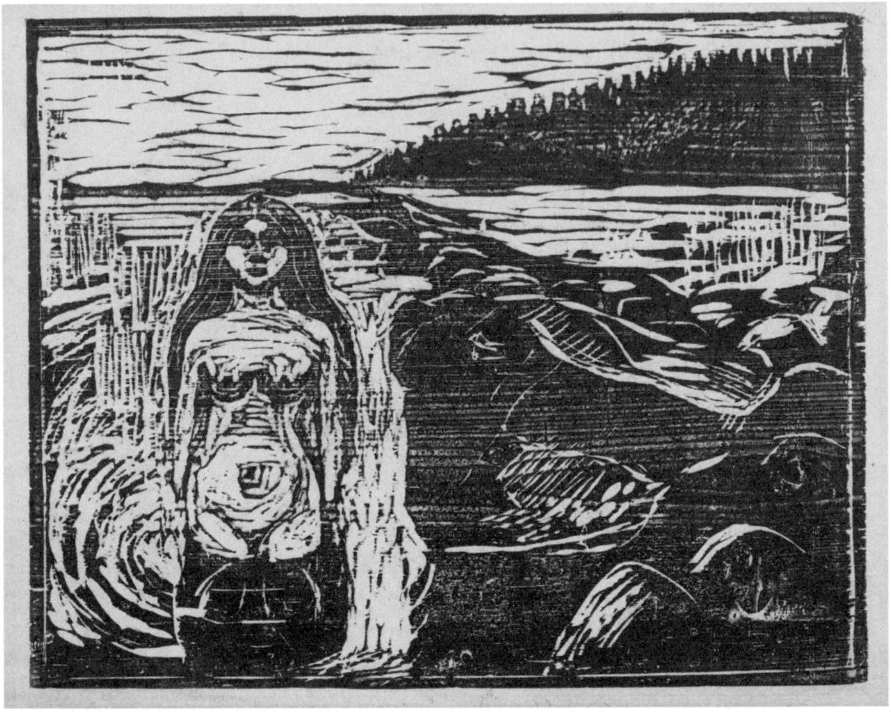

Edvard Munch. *Badende kvinne—Woman Bathing*. 1899. Woodcut. Image: 42 × 51.8 cm. Sheet: 44.4 x 58.9 cm. Munch Museum, Oslo. *Photo* © Munchmuseet

Munch did not stop there his research concerning the stratification of images or layers in one and the same picture. For example, in his woodcut production, the interaction between the represented subject and the wood grain is very often sought, as if to emphasise, through the interference, the potential and the physical constraints of creating an image, the genetic emergence of the work of art from its medium. In *Woman Bathing* (1899), which in terms of relation between woman and nature looks iconologically close to the painting analysed above, the female figure, just like the landscape, is permeated by the texture of the wood and, in height, strangely fits into the lake.[18] The wood grain contributes to blur the figuration and provides an uncanny aspect to her statutory position, almost buried in the water. About fifteen years after *From Nordstrand*, Munch pursued the stratification of layers in another aesthetic medium: photography. Through the use of long exposure or the double exposure technique, he began to reproduce first his own image (see *Self-portrait*, 1907, and *Self-portrait with Housekeeper at 53 Am Strom, Warnemünde*, 1907), principal subject of his photographic production, then some women of his entourage (*Rosa Meissner*, 1907, and *Nurse in Black. Dr. Jacobson's Clinic in Copenhagen*, 1908–1909). All these shots, called by him a year before his death "the photos of destiny",[19] show

[18]Edvard Munch made another woodcut of the *Woman Bathing* in colour (1899).

[19]The denomination was given by Munch in a 1943 annotation. Cf. Meyer (2005, p. 242).

fuzzy, evanescent human figures, disappearing in transparency behind the solidity of the objects or within the steadiness of the photographed environments. Losing henceforth the archetypal connotation that could still be found in the woman of *From Nordstrand*, these figures nonetheless act as similar traces of ghostly presences, absorbed by the picture in layers.

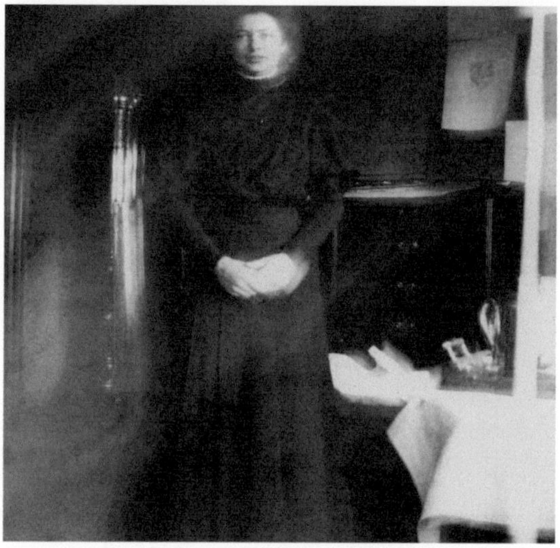

Edvard Munch. *Sykesøster i sort, Dr. Jacobsons klinikk—Nurse in Black. Dr. Jacobson's Clinic in Copenhagen.* 1908–1909. Photograph. The Munch Museum. *Photo* © Munchmuseet

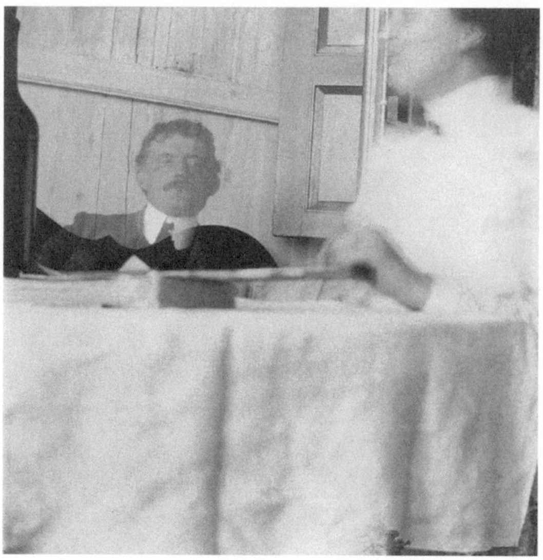

Edvard Munch. *Self-portrait with Housekeeper at 53 Am Strom, Warnemünde*. 1907. Photograph. Image: 8.7 × 8.4 cm. Sheet: 8.9 × 8.9 cm. The Munch Museum. *Photo* © Munchmuseet

The exposure in differentiated times underlines, more than in painting, their ephemeral presence, the diaphanous appearance of their existence. If these pictures belong to Munch's project *to photograph the soul*, in my opinion, it is in the sense that they participate in his will to fix the potential of life and therefore create an art *crystal*. If in the photos the work of the figurable generates a perception more enigmatic than the signifying load of interpretations produced in *From Nordstrand*, this is perhaps finally because, according to the artist's point of view:

> The camera cannot compete with painting as long as it cannot be used in heaven or hell. (Munch in Tøjner (ed.), 2001, p. 72)

References

Caliandro, Stefania: L'art comme cristallisation de la vie: réflexions sur une œuvre d'Edvard Munch (*Paysage à Ljan. Désir brûlant*, 1892). Bulletin de la Société française d'esthétique. Paris. pp. 1–12 (2007)

Damisch, Hubert: Narcisse Baroque? In: Christine, B. (eds.) Puissance du Baroque. Les forces, les formes, les rationalités. Éditions Galilée, Paris, pp. 29–42 (1996)

Gamboni, Dario: Potential images: ambiguity and indeterminacy in modern art. Reaktion, London (2002)

Meyer, Robert: Le fotografie del destino e autoritratti. Su Edvard Munch e i suoi ritratti fotografici. In: Bjerke, Ø.S.; Bonito Oliva, A. (eds.) Munch 1863–1944. Skira, Milanpp, pp. 241–243 (2005)

Bjerke, Øivind Storm; Bonito Oliva, Achille (eds.): Munch 1863–1944. Catalogue of the exhibition at the Complesso del Vittoriano in Rome from 10 March to 19 June 2005. Skira, Milan (2005)

Bjerke, Øivind Storm: Il maestro del "non finito". In: Bjerke, Ø.S., Oliva, A.B. (eds.) Munch 1863–1944. Skira, Milan, pp. 21–39 (2005)

Tøjner, Poul Erik (ed.): Munch in his own words [original: Munch—Med egne ord]. Prestel Verlag, Munich (2001)

Woll, Gerd (ed.): Edvard Munch samlede malerier. Catalogue raisonné. 4 vols, Oslo (2008) [English transl.: Munch, E.: Complete paintings. Thames & Hudson Ltd., London (2009)]

Chapter 9
Feeling the Complexity, Saying the Inconsistent

In his diaries, on 30, June 1902, when his artistic career was still finding its earliest foundations, Paul Klee wrote there were at that time three things: *physis*, of which the Greco-Roman antiquity made him think, with its "objective attitude, worldly orientation and architectonic center of gravity"; *psyche*, related to a Christianity "with a subjective attitude, other-worldly orientation and musical center of gravity". "The third–Klee concludes–is the state of the modest, ignorant, self-taught man, a tiny ego." (Klee 1957, English transl.: pp. 125–126). What does it mean this way of placing himself as a third term between physis and psyche, to do it as an autodidact, after leaving the academy and having been marked by a stay in Italy and Rome?[1] To learn all over again (cf. Klee 1957, English transl.: p. 124): to wonder about the (gravitational) architectonic of plasticity and, at the same time, capture in it the subjectivity of a transcendent orientation, such as he felt in music.

Say the sensitive: say, express, write in painting and, on the other hand, let the painting be said, described. The relationship of art with the logos, or of the image with words, is an old issue, perhaps inexhaustible, which the work of Michel Foucault also bears witness to. In *Ceci n'est pas une pipe*, despite the principle of subordination between text and image that Foucault has just defined as necessary ("In one way or another, subordination is required. Either the text is ruled by the image [...]; or else the image is ruled by the text"), the author observes that Klee has abolished the sovereignty of this principle (Foucault 1973, pp. 39–45, esp. 39–41; English transl.: pp. 32–35, esp. 32–33), interwoven verbal signs and spatial figures (cf. ibid., pp. 48

[1]This question, as well as the present ensuing study, have been raised by the paper of Jean-Claude Coquet, "La sémiotique visuelle permet-elle d'accéder à la «réalité»?" (Does visual semiotics make it possible to access "reality"?), read on 15 April 2010 at the Congress of the International Association for Visual Semiotics in Venice (Coquet 2011). The spelling *physis*, instead of *phusis* given by Coquet, is here retained in reference to Klee's text. Cf. also the thematic issue "Phénoménologie et sémiotique" of the journal *Actes sémiotiques*, 114, Université de Limoges, 2011 (available at http://epublications.unilim.fr/revues/as/2730) with "Le débat", introduced by Ivan Darrault-Harris, between Jean-Claude Coquet and Jean Petitot. The original French version of my essay, "Sentir la complexité, dire l'inconsistant", has been published in the academic journal *Littérature* (Caliandro 2011).

S. Caliandro, *Morphodynamics in Aesthetics*, Lecture Notes in Morphogenesis,
https://doi.org/10.1007/978-3-030-29631-5_9

and 53; English transl.: pp. 36–37 and 39), suppressed the hierarchical order running from form to discourse or from discourse to form. But the philosopher does not address the question of the logos of art to which some lines, actually a little elliptical, of Louis Marin incite us. About Klee, Marin speaks of a "structural painting – or better, a structural analysis model of painting – conditionally upon understanding that, for Klee, the structure is only the trace of a force, the trail left by an energy, the play of an elementary activity" (Marin [1970] 1971a, p. 104, own translation). The semiologist's suggestion goes beyond the idea of a simple hybridisation between painting and writing (cf. Marin [1968] 1971b, p. 97) and reaches the point of the feeling, of how to make the feeling *visible*.

I would like to resume these thoughts concerning the saying of art, its logos, its faculty to say what one feels, as well as its potential to be said, even to be somehow described, on the basis of the work of two artists who were hugely concerned with this issue in quite different ways: on the one hand, Paul Klee with his thinking by images, on the other hand, Marcel Duchamp whose art defies its own figuration. Moreover, the two constantly cross plastic expression with its discursive transposition in language, in particular, Klee in his lectures and courses at the Bauhaus, giving voice to his plastic thinking, Duchamp in his propaedeutic notes and in those descriptive of his art.

Polyphonic Vibration of Differences

> What does "nature" count for here? The real point is the law according to which "nature" functions and how it is revealed to the artist.
>
> (P. Klee, *Diaries*, 1905)[2]

> He has so much to say that a Klee is never like another Klee.
>
> (M. Duchamp, 1949)[3]

Louis Marin carefully sketches the idea that Klee has developed a "structural analysis model of painting", since the artist's writings nuance each of these terms (*analysis*, *structure*), confront, juxtapose and even combine each one with its opposite. In addition to deepening his knowledge of painting, more than other forms of art, for simple personal predilection, and not favouring the study of art over the observation of nature, Klee affirms on several occasions that the analytic way ensues from

[2]Klee (1957, English transl.: p. 185).

[3]Marcel Duchamp. Klee, 1879–1940. Painter, Graphic Artist, Writer. In: George Head (ed.). From the Catalog. Collection of the Société Anonyme (Critical studies written from 1943 to 1949). 1950. New edition: Michel Sanouillet (ed.) (1994, p. 204; English transl.: p. 151).

the limits of language, especially when the latter aims at the clarity of the didactic argumentation:

> It is not easy to orient yourself in a whole that is made up of parts belonging to different dimensions. And nature is such a whole, just like art, its transformed reflection.
>
> It is hard to gain an overall view of such totality, whether it is nature or art, and it is still harder to communicate the view to others.
>
> The answer lies in methods of handling spatial representation which lead to an image that is plastically clear. The difficulty lies in the temporal deficiency of language.
>
> For in language there is no way of seeing many dimensions at once. (Klee 1956, English transl.: pp. 82 and 86)[4]

The "temporal deficiency of language" consists in missing the phenomenon of simultaneity "in the realm of verbal explanation". It has been "eloquently" overcome in music, temporal art too, thanks to "polyphony". Indeed, "this simultaneous view of many dimensions" is, according to Klee, "what the so-called spatial arts have long succeeded in accomplishing", painting included (ibid.).[5]

The complexity of feeling, both in nature and art, has been exemplified by the imaginary little journey the artist narrates by using "a topographical plan":

> The first act of movement (line) takes us far beyond the dead point. After a short while, we stop to get our breath (interrupted line or, if we stop several times, an articulated line). And now a glance back to see how far we have come (countermovement). We consider the road in this direction and in that (bundles of lines). A river is in the way, we use a boat (wavy motion). Farther upstream we should have found a bridge (series of arches). (Klee 1956, English transl.: p. 76)[6]

To express in language what he perceives in crossing a landscape, Klee does not hesitate to associate to it visual elements that synthesise things, movements as well as moods. The journey continues: the delightfulness of meeting a companion becomes a "(convergence)", then "(two separate lines are drawn)"; "A certain agitation" is rendered by: "(expression, dynamics, and psyche of the line)"; "an unploughed field": "(area traversed by lines)"; "fog": "(spatial element)"; some basket weavers "returning home with their carts": "(the wheel)"; "a child with merriest curls": "(spiral movement)"; a "flash of lightening on the horizon": "(zigzag line)"; stars over us: "(field of points)" (ibid.). The evocation of graphic and topological elements, written down in parentheses, punctuates the impressions of which "a short trip of this kind" is full and which can later come back to mind in "a number of memories" free of any chronological constraint (ibid.).

---

[4]From the lecture Paul Klee delivered in Jena on 26 January 1924 (Klee 1956, English transl.: pp. 81–95).

[5]Cf. Caliandro (2001–2002).

[6]From Paul Klee's "Creative Credo", first published in 1920 (Klee 1956, English transl.: pp. 76–80).

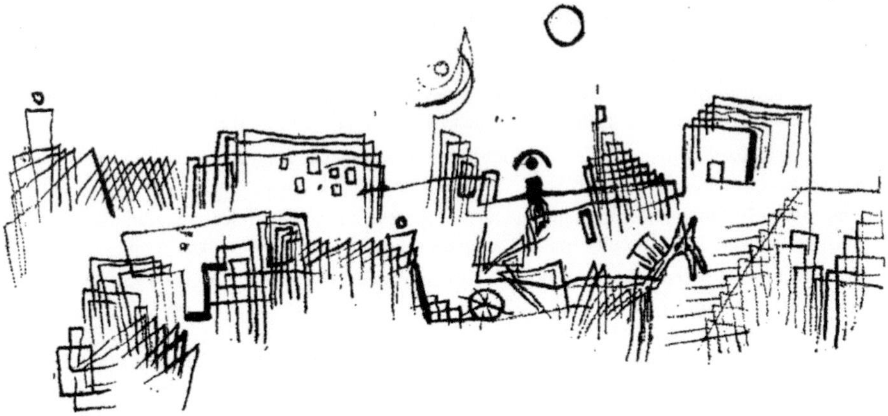

Paul Klee. *Schwere Fahrt durch O—Difficult journey through O.* 1927, 139. Pen on paper on cardboard. 15 × 30.5 cm. Swiss private collection, deposited in Zentrum Paul Klee, Bern. *Photo* © Zentrum Paul Klee

By comparing this tale with the drawing Klee made in 1927, *Difficult journey through O.*, one can understand how the linearity of the strokes is enriched, in the picture, by the simultaneity of the moments represented which, thence, echo one another, concatenate and merge. In fact, if the tremor, the oscillation, the outpouring, the breaking or the overlapping of a line suffices to convey a certain vibration, if a line carries an expression, a dynamism and, even, a state of mind, the relationship of complementarity, resonance or conflict that the line engenders with the other elements of the work of art prompt us not to forget the complexity of which both nature and art are composed. Each part or element assumes a *measure*, a *weight* and a *quality* in relation to the other parts or elements. The analytic approach is, therefore, just a way to shed light on the *synthesis of differences* which, according to Klee, constitutes any higher organism (cf. Klee 1956, English transl.: pp. 429, 433, 444–446).

Just as the analysis serves as a heuristic tool to approach the polyphony of differences, the notion of structure must also be handled with caution. Let's consider the attention Klee draws to the leaf. The appearance of a line is able to visualise the different figures the form of a leaf can have: "Saw-toothed", "Serrate", "Scalloped", "Dentate", "Sinuate", "Fringed", "Fretted" (Klee 1970, English transl.: p. 16). The graphic design that Klee employs to delimit the phytomorphic contour is, however, the outcome of a thorough study of the morphogenesis by which the leaf takes form (cf. ibid., pp. 3–23). He analyses the structure of the veins, their layout and branching until their progressive decrease no longer allows distinguishing between "linear forces" and "planar elements" (ibid., p. 7). Thus, the "constructive, articulating forces" conveyed by the veins lead to think of "the evolution of the leaf (in the pictorial sense)" into the "two-dimensional massiveness":

> The planar massiveness is the element that to the eye no longer appears linear, but it is distinguished as a separated element by its tangle of lines. As against linear definition, this element may create an impression of softness.

> The line system reaches into the other element, finely forked or sievelike, and enough of it sticks.
>
> The planar form that arises is then independent of the inreaching linear radiation.
>
> And where linear power ends, there arises contour, the limit of planar form. (Ibid., p. 13)

Also, if the radii of the veins push beyond certain dimensions of the edge, the latter is hollowed out and rips in the opposite direction, as it happens in many types of leaves (cf. ibid., p. 21). The analysis focuses on describing the morphology of the object insofar as it keeps track of "formational and at once articulating" energies (ibid., p. 25). The living form manifests the forces that modelled it, it contains its own formation; in the same way, art is the impression in the matter of a creative "dictate" (or law), it makes matter animated (ibid., p. 45).

The example of the thin plate covered with fine sand, that a violin bow makes vibrate, explains Paul Klee's idea of structure: "This vibratory impulse is the heart of the matter. It causes the sand to arrange itself in a corresponding rhythmical order." (Ibid., p. 44). The moving flow, thus drawn by the oscillation of the sand particles, is the expression of the vibration that models them. The structuring law is then not only visualised in lines of force, which are more obvious, such as the veins of the leaves, but also crosses and sets a rhythm to the whole matter, even when its propagation becomes so diffuse and minute that it is not perceivable to human sensitivity (cf. ibid., p. 51). Hence, the vibration of the creative force does not concern the only structure, which, moreover, is not essential in itself and may on the contrary be liable to "an alternation of distinct and blurred structures" or even "an alternation of structure and absence of structure" (ibid., p. 219).[7] This vibration spreads from the structure to the whole and endows the form with a function that animates the work (whether the latter be natural or artistic).

The evolution from an element to its opposite—from the structure to the absence of structure, but that also applies to the plastic values of the line, the light and shade and the colour, with their polarities—is not necessarily carried out through opposition but can pursue scaling, nuancing or interpenetrating effects. The approach proceeding by the division of elements, taken separately, undoubtedly encounters, according to Klee, "the problems that necessarily stand in the way of an analytical approach" (ibid., p. 281).[8] Nevertheless, it is suitable to always consider "by way of contrast" the proportions that govern the structure, for it is from these proportions that "form-determination issues" (ibid., p. 283). Thus, sets of weights and of counterbalances of values, at once graphic, chromatic and luminous, give to the painting architectonics its dynamism and several kinds of equilibrium in composition (besides static equilibrium, for instance, unstable equilibrium and asymmetrical balance)

---

[7] Constantly reflecting on many architectonic variants and their effects, Klee also considers "[s]cattered" and "compact" structures; (cf. also Klee 1970, English transl.: p. 217).

[8] Cf. also the observations on the "fine flow from one pole to the other", for instance "from white to black" (ibid., p. 347), and how he suggests to paint the variations between: "Standing", "Gliding", "Striding, small strides", "large strides", "Leaping in small strides", "Leaping in jumps". "Gliding is soft (fluid, gaseous)", whilst, according to him, walking and jumping (which are harder) are solid (ibid., p. 349).

(cf. Klee 1956, English transl.: pp. 197–214). Klee teaches to study effects resulting from these contrasts, which are stronger as regards basic contrasts and more attenuated as for secondary ones, not forgetting forms of mediation that make composition even richer (cf. ibid., p. 319). In this way, the artist constructs an instrument of analysis of the complexity that he feels in art and nature and which, in his terms, goes beyond theory. He imagines a balance of forms, of light and shade and of colours, which he proposes to dynamise even more in a pendulum, a pendulum with a mobile suspension point and a spinning top (cf. ibid., pp. 385–393). He also examines the spatial complexity in forms of interpenetration of inner and outer spaces, of combination of different perspectives, such as projections on plane, front and side views or irregular projections on an uneven surface, in simultaneous multi-dimensional phenomena, in slight or pronounced deviations from the normal paths of perspective construction as well as other anomalies with regard to rules hence surmountable, up to the synthesis of several viewpoints and the conception of shifting viewpoints (cf. ibid., pp. 154–159, 173–175 and 253). In this sense, the analytic way seems to him the only possible way to *speak* of the coexistence of the various aspects and dimensions perceived. It is a functional and instrumental way to identify a dictate and measure the deviations from it in the work of art.

His plastic thinking transcends then the phenomenal physics of the object, aims at the architectonics of the object and attains the function by which the object takes form. Feeling the complexity means coming into vibration with the forces that are at work and manifest themselves in the work. Klee's intention in respect to nature differs markedly from the Impressionists' research on the object appearance, "the object's surface filtered by the air" or, again, the photographic rendering of the world appearance (ibid., p. 63[9]). Likewise, he distances himself from an abstract aim, in the narrow sense, since, even when creating an imaginary world, Klee seeks to ensure the latter remains "credible", "in the realm of the human" (ibid., p. 262). From his point of view, contradictions of this kind are not unfavourable because the phenomenal world and the abstract world, the *organic* connection and the *elements that can be read abstractly* can meet, as he affirms to happen in memory (cf. ibid., 309). It is finally not useless to recall the denomination he gave himself in 1915. In contrast to a phenomenological art centred on purely optical impressions, also wary of an abstract formalism leaning towards sterile aridity, Paul Klee emblematically defined himself as "abstract with memories" (Klee 1957, English transl.: p. 315).

> The object grows beyond its appearance through our knowledge of its inner being, through the knowledge that the thing is more than its outward aspect suggests. [...] This is visible penetration, to some extent that of a simple knife, to some extent helped by finer instruments which make the material structure or material function clear to us. (Klee 1956, English transl.: p. 66)

In his commitment to both grasping and transcending the surface of things, Klee manages to elucidate plastically the—semiotic—function, to make manifest "visible

[9]From Paul Klee's *Wege des Naturstudiums* ("Ways of Nature Study"), first printed in 1919–23 and published in 1923 (Klee 1956, English transl.: pp. 63–67).

penetration", viz. the visualisation of forces that, in the form, determine its morphology. To understand the logos of art is hence to perceive how, between *physis* and *psyche*, the creative mind goes through matter, animates it according to a plastic architectonics and impresses in it a polyphonic vibration, which continues to resonate in the work of art since its origin.

Metalogic of the Possible, or Saying the Logos of Art

> I mean the difference between speaking about red and looking at red.
>
> (M. Duchamp, 1965)[10]

> Metalogic is concerned with the smile, the gaze, the scent, all the seductions between good and evil.
>
> (P. Klee)[11]

The art of Marcel Duchamp also departs from the phenomenal viewpoint of the Impressionist movement, since, as the artist claims in retrospect, he "wanted to get away from the physical aspect of painting"; and he insists, he was interested "in getting away from the physicality of painting" "to put painting once again at the service of the mind" (Duchamp in Sanouillet and Peterson (eds.) 1973, p. 125).[12] Though this opened him to accusations of intellectualism, he, nonetheless, judged Dadaism "very serviceable as a purgative" for its "metaphysical attitude" generating "a sort of nihilism" (ibid.).[13] Marcel Duchamp's personal approach does not boil down, however, to this moment of protest. A concern of "philosophical outlook" made him think of art "on a broader scale", in relation to the discussions, at that time, on the fourth dimension and the non-Euclidean geometry (ibid., p. 126). "A reaction of visual indifference" leads him, on the one hand, to seek the "complete anesthesia" of the sensory aspect of the work of art, especially through readymades (ibid., p. 141),[14] on the other hand, to further research on "*a certain inopticity*" of

---

[10]Marcel Duchamp, note added to "À l'infinitif «Boîte blanche»", in: Sanouillet (ed.) (1994, p. 118).

[11]Klee (1956, English transl.: p. 60).

[12]From Marcel Duchamp's "The Great Trouble with Art in This Country", remarks collected in English in 1949; Sanouillet and Peterson (eds.) (1973, pp. 123–126).

[13]The idea of the dada movement as a purge has also been used by Hans Arp in its definition in 1925: "The dadaïsm has assailed fine-arts. He declared art to be a magic purge gave the clyster to Venus of Milo and allowed «Laocoon and Sons» to absent themselves at last after they had tortured themselves in the millennial fight with the rattlesnake. Dadaïsme has carried affirmation and negation up to nonsens. In order to come to the indifference dadaïsme was distructive." Hans Arp in: Lissitzky and Arp (eds.) (1925, p. X) (the quote is given here in its original English spelling).

[14]From M. Duchamp's "Apropos of «Readymades»", talk delivered in 1961 and first published in English in 1966; new edition in: Sanouillet and Peterson (eds.) (1973, pp. 141–142) (all the text is in upper case).

art, of which *The Large Glass* is, in a way, his major upshot, despite (or thanks to) its incompletion (ibid., p. 83).[15] He focuses on the idea of a representation that refers to "a *reality which would be possible by slightly distending* the laws of physics and chemistry" (ibid., p. 71).[16]

His thoughts on the *possible*[17] orient, in fact, the design of *The Large Glass* wherein are projected, according to the author, forms of dimension greater than the third. Thus, he plans to make "a painting on glass so that it has neither front, nor back; [...] (to use probably as *a three-dimensional physical medium* in a four-dimensional perspective)" (Duchamp 1999, p. 41; English transl.: p. 79).[18] Furthermore, he points out it will be a "Possible", "Rather hypophysics" instead of physics, properly speaking (ibid., p. 51; English transl.: p. 153). Duchamp's affirmation that the Bride "is the bi-dimensional representation of a tri-dimensional Bride who is herself the projection of the quadri-dimensional Bride into the tri-dimensional world" leads then Jean-François Lyotard to understand that *The Large Glass* "figures the unfigurable; [...] the shadow cast, on its plan, by a figure that could not be intuited" (Lyotard 1977, new edition: pp. 114 and 112; English version: pp. 115 and 113 respectively). I will not include here the details of this superb reading given by Lyotard in his collected essays *Les Transformateurs Duchamp*, coupling the analysis of *The Large Glass* and that of *Étant donnés: 1° la chute d'eau, 2° le gaz d'éclairage...* (*Given: 1. The Waterfall, 2. The Illuminating Gas...*) with the writings of the artist until then published. The "[h]inge logic", which concerns both the internal articulation of Duchamp's works of art and the articulation of these two works with each other, reveals the "mirrorish" aspect (evoked "against the specular") of forms projected in a "dissimilating" way—following both the fidelity and the infidelity of the mirror (ibid., new edition: pp. 140, 116 and 80; English version: pp. 141, 117 and 81 respectively). And for Lyotard, it refers to this "curious property" that Kant names "incongruence", according to which the right hand and the left hand are similar and symmetrical but "non-superimposable" on each other (ibid., new edition: p. 86; English version: p. 87). I would then like to suggest that this same hinge logic, this mirrorish metamorphosis, is finally what articulates, in the art of Duchamp as well as in art in

---

[15]From M. Duchamp's "À l'infinitif «Boîte blanche"», in: Sanouillet (ed.) (1994, pp. 105–141); English transl.: Sanouillet and Peterson (eds.) (1973, pp. 74–101). Concerning this inopticity the artist mentions: "The colors about which one speaks." and adds, later, the note I put in exergue of this subchapter. The artist worked on *The Bride Stripped Bare by Her Bachelors, Even*, also known as *The Large Glass*, from 1915 to 1923, when he finally gave up going ahead with this project.

[16]From M. Duchamp's "La «Boîte verte»", notes made by the artist in 1934 and printed in three hundred copies (Sanouillet (ed.) 1994, pp. 39–102; English transl. in: Sanouillet and Peterson (eds.) 1973, pp. 26–71). In this quote, Duchamp refers to both the physics and the chemistry of spatial forms (e.g. distribution of molecules, roundness and flatness), colours and light sources (their matter).

[17]"The figuration of a possible. / (not as the opposite of impossible / nor as related to probable / nor as subordinated to likely) / the *possible* is only / a *physical "caustic"* [vitriol type] / burning up all aesthetics and callistics". Marcel Duchamp, "Possible", [1913] published in 1958; new edition in: Sanouillet (ed.) (1994, p. 104; English transl.) in: Sanouillet and Peterson (eds.) (1973, p. 73). Le symbol / indicates the end of line.

[18]From Marcel Duchamp's "Le grand verre" in: Duchamp (1999, pp. 37–102).

general, the saying of the work of art in relation to its possible complement, namely the possibility for the work of art to be said.

In the preamble of "Hinges", Lyotard argues that this hinge logic meets with "the enunciation of meta-non-sense, or the contingency of the law, [...] itself regulated by an inclusive disjunction" (ibid., new edition: p. 140, English version: p. 141).[19] The "ironism" Duchamp makes use of—at least this "ironism of affirmation" which differs "from the negating ironism, depending only on the Laugh" (M. Duchamp quoted ibid., new edition: p. 138; English version: p. 139)—proceeds without any logical relation of implication[20]; more precisely, as Lyotard remarks, it "stand[s] in the way of one of the great operators of congruence, for example implication (if *p*, then *q*), which is the very serious logical causality" (ibid., new edition: p. 146; English version: p. 147).[21] It is not, therefore, this serious logical causality but an "ironic causality" that, in the coupling of Duchamp's writings and his works of art, motivates the choice between several possible realisations (M. Duchamp quoted ibid., new edition: p. 138; English version: p. 139). Lyotard states:

> To justify the choice is to transform contingency into necessity or permission into obligation; it's to pass from the "it's possible" to "it's necessary". To choose the justification (of the choice) is to reintroduce contingency into the enunciation of necessity: "it is possible that it would be necessary"; or the permission into that of obligation: "it's permitted that one must". You transform non-sense into sense, and this sense is itself hung from a non-sense. (Ibid., new edition: p. 140; English version: p. 141)

This mirrorish connection between aesthetic thoughts and plastic production, between texts and works of art, occurs not only in the genesis of the creation but, as it was in a symmetrically reversed manner, also at the time of its exegesis. The ironism of Duchamp's art affects the act of interpretation, and the philosopher is well aware of this when, "beyond truth-values", he speaks of "inability and power" (ibid., new edition: p. 96; English version: p. 97)[22]:

> Innocent is he who thinks you can stand in inconsistency as you might stand in seriousness: i.e. consistently. You must hold yourself inconsistently in inconsistence, and mix segments of consistency in with it, and make them indiscernible from the others. (Ibid., new edition: p. 66; English version: p. 67)

Ironism, "affecting any enunciation", leads to the development of an inconsistent saying, wherein the inconsistency "is not insignificance", but a precious non-sense one must/can cultivate (ibid., new edition: pp. 142 and 56; English version: pp. 143 and 57 respectively). The incongruence between figures and between spaces (cf. ibid.,

---

[19]The inclusive disjunction corresponds to the linguistic expression "and/or". Lyotard (1977, new edition: p. 146; English version: p. 147).

[20]Cf. Herman Parret, "Preface", ibid., new edition: pp. 10–23, esp. 21; English version: pp. 32–45, esp. 43.

[21]Lyotard also point out that "Duchamp's machines [...] know no consequence", they sketch an incomplete implication in which the involved statement lacks, and their language "consists in positing a state of fact as if you were going to draw some consequences from it, and then not drawing them". Ibid., new edition: p. 100; English version: p. 101.

[22]Lyotard's text, "Machinations" [1974], begins emblematically with the sentence: "To interpret is futile." Ibid., new edition: pp. 96–138, esp. 96; English version: pp. 97–139, esp. 97.

new edition: p. 92; English version: p. 93), with its discrepancy between *appearance* and *apparition*, that is to say between the "sensory evidence enabling one to have an ordinary perception" of the object and "the *mold* of the *object*" which determines its "mirror image" (Duchamp in Sanouillet (ed.) 1994, pp. 120–121; English transl.: pp. 84–85),[23] thus does not concern only the complex spatial matters the artist considers regarding the transposition into different dimensions, but it also touches on the definition of the *dictate* according to which, to use Klee's terms, art *functions*.

To think of art "on a broader scale", as Duchamp wrote (see *supra*), implies this transformative performance, this projection of art to another level of logos that Lyotard calls "*transformance*" (Lyotard 1977, new edition: p. 70; English version: p. 71). It is a metalogical enunciation which, without being dictated by a logical implication, strictly speaking, does stand in the incongruence, in the *possible* passage between one semiotic space and the other, in the reflective symmetry of the figures, not superimposable, of feeling and saying. To introduce an art of saying into plastic creation, as Duchamp does, means not only to include the figurable in it but also to hinder figuration. And yet the artist seems to indicate a fine gap between, on the one hand, his way of conceiving the saying inside his art and, on the other hand, the possible transposition of art into interpretation. In the first case, the possible gives rise to a passage, an *infra-slim* metamorphic transformation[24]; in the second it allows interpretation "to rave as one wishes".[25] The commentary, as any saying about art, involves the identification of a dictate that is supposed to reflect the semiotic function of the work of art but is rather its glint due to topological projection. As Lyotard still argues: "Any putting into perspective, including that of our [...] optic, is the imposition of an order based on *faits divers*, the interesting thing is that this order has no reason or principle." (Lyotard 1977, new edition: p. 206; English version: p. 207).

As discussed previously, following a suggestion of Louis Marin, Paul Klee's plastic thinking invites to feel the logos of art. Conversely, an art of logos is configured with various modes in Marcel Duchamp's art as well as in Jean-François Lyotard's subsequent philosophical observations. Indeed, Duchamp defies the sensuous nature of art through the discourse he installs in it; Lyotard consciously stands in the incongruence of a saying that develops by interpretation. These are different ways of saying the sensitive, intertwining at various levels the discourse and what one feels. Plas-

[23]From M. Duchamp's "À l'Infinitif" in: Sanouillet (ed.) (1994, pp. 105–141; English transl.: pp. 74–101). With regard to the concept of mould, Duchamp gives this "typical" example of "*Moulds in folds*": "trousers worn and very marked by folds. (giving a sculptural expression of the individual who wore them)". He then specifies the technical reversal that occurs, in comparison to the manual execution of an original sculpture: "by wearing the trousers, the leg works like the hand of a sculptor and produces a mould (instead of moulding), and a fabric mould that expresses itself in folds". From M. Duchamp's "Inframince" in: Duchamp (1999, pp. 21–36, esp. 34, own translation).

[24]As well as other texts now published, Duchamp's notes on the infra-slim were not known by Lyotard (cf. Duchamp, "Inframince", in: Duchamp 1999, pp. 19–36). However, a sentence already illustrated this notion: "When / the tobacco smoke / also smells / of the mouth / which exhales it / the two odors / are married by infra-slim"; Duchamp in Sanouillet (ed.) 1994, p. 274; English transl. in: Sanouillet and Peterson (eds.) 1973, p. 194 (all the quote is uppercase).

[25]Marcel Duchamp, letter on 4 October 1954. In: Sanouillet (ed.) (1994, p. 235, own translation).

tic semiotics and discursive semiotics intersect and then reflect one another without overlapping, if not in the sense of a Hjelmslevian catalysis transposing a function into the meta-function that glosses it. The scientific stance of this semiotic interlinking maintains, in the honesty of its own discourse, the consciousness of an interpretative development based on the possible. Finally, let me rephrase one of the Lyotard's statements about Duchamp, extend it to art in general and abandon his factitious care of "hid[ing] the one important thing", namely that we are interested in art in inverse proportion to the amount we have understood about it (ibid., new edition: p. 54; English version: p. 55).

References

Caliandro, Stefania: Le point gris de Paul Klee. Visio, 6, 4, winter, Canada, pp. 21-41 (2001–2002)

Caliandro, Stefania: Sentir la complexité, dire l'inconsistant. Littérature, 163, 20–31, Sept, Paris, pp. 20-31 (2011). Also at: https://www.jstor.org/stable/i40080124

Coquet, Jean-Claude: La sémiotique visuelle permet-elle d'accéder à la «réalité»? In: Migliore, T. (ed.) Retorica del visibile. Strategie dell'immagine tra significazione e comunicazione. 1. Conferenze. Aracne Editrice, Rome, pp. 111–120 (2011)

Duchamp, Marcel: Notes. Flammarion, Paris (1999) [English transl.: Adcock, C.E. (ed.) Marcel Duchamp's Notes from the Large Glass: An N-dimensional Analysis. UMI Research Press (1983)]

Foucault, Michel: Ceci n'est pas une pipe. Fata Morgana, Montpellier (1973) [English transl.: This is Not a Pipe. With Illustrations and Letters by René Magritte. University of California Press, Berkeley, Los Angeles, London (1983)]

Foucault, Michel: Klee, Kandinsky, Magritte. Ceci n'est pas une pipe. Fata Morgana, Montpellier, pp. 39–45 (1973) [English transl.: Klee, Kandinsky, Magritte. This is Not a Pipe. With Illustrations and Letters by René Magritte. University of California Press, Berkeley, Los Angeles, London, pp. 32–35 (1983)]

Klee, Paul: Das bildnerische Denken, collected and edited by Jürg Spiller. Schwabe & Co, Verlag, Basel, Stuttgart (1956) [English transl.: Notebooks. Volume 1. The Thinking Eye. Lund Humphries, London (1961)]

Klee, Paul: Tagebücher von Paul Klee 1898–1918, edited by Felix Klee. Verlag M. DuMont Schauberg, Cologne (1957) [English transl. authorized by Felix Klee: The Diaries of Paul Klee, 1898–1918. University of California Press, Berkeley, Los Angeles, London (1964)]

Klee, Paul: Unendliche Naturgeschichte, collected and edited by Jürg Spiller. Schwabe & Co., Verlag, Basel, Stuttgart (1970) [English transl. Notebooks. Volume 2. The Nature of Nature. Lund Humphries, London (1973)]

Lissitzky, El; Arp, Hans (eds.): Die Kunstismen. Les ismes de l'art. The Isms of Art. Eugen Rentsch Verlag, Erlenbach-Zurich, Munich, Leipzig (1925)

Lyotard, Jean-François: Les Transformateurs Duchamp. Galilée, Paris (1977) [New bilingual (French-English) edition: Les Transformateurs Duchamp. Duchamp's TRANS/formers. Van Looveren & Princen, Brussels (2010)]

Marin, Louis: Klee ou le retour à l'origine. In: Études sémiologiques. Écritures, Peintures. Klincksieck, Paris, pp. 101–108 ([1970] 1971a)

Marin, Louis: Comment lire un tableau. In: Études sémiologiques. Écritures, Peintures. Klincksieck, Paris, pp. 89–99 ([1968] 1971b)

Sanouillet, Michel; Peterson, Elmer (eds.): The Essential Writings of Marcel Duchamp. Oxford University Press, New York (1973)

Sanouillet, Michel (ed.): Duchamp du signe. Écrits. Flammarion, Paris (1994) [English transl.: Sanouillet, M., Peterson, E. (eds.) The Essential Writings of Marcel Duchamp. Oxford University Press, New York (1973)]

MIX
Papier aus verantwortungsvollen Quellen
Paper from responsible sources
FSC® C105338

If you have any concerns about our products,
you can contact us on
ProductSafety@springernature.com

In case Publisher is established outside the EU,
the EU authorized representative is:
Springer Nature Customer Service Center GmbH
Europaplatz 3, 69115 Heidelberg, Germany

Printed by Libri Plureos GmbH
in Hamburg, Germany